Pi,
Monads,
and the
Quasi-Circle Theory

Pi,
Monads,
and the
Quasi-Circle Theory

**A theory on the circle more appropriate
to the space age**

Lionel Fabius

To order additional copies of this book, contact:
Xlibris Corporation
1-888-795-4274
www.Xlibris.com
Orders@Xlibris.com
76249

"I praise you, Father, Lord of heaven and earth, because you have **hidden** these things from the **wise** and learned, and revealed them to little children. Yes, Father, for this was your good pleasure." Luke 10:21

A Step Beyond Pi

Introduction to the Quasi-Circle Theory

A circle has a geometrical and a numerical aspect. This work is the study of the circle seen from a numerical point of view.

For the past two thousand years, no significant progress has been made to improve the methods used in the calculation of circles. Due to the transcendence of pi, the inner and outer dimensions of the circle were never calculated with precision, only approximately. We were never able to calculate the number of squares that makes the area of a circle in the same manner that we calculate the area of a square or a rectangle; yet, the area of a circle is expressed in terms of square units. So far, the perimeter of a circle cannot be determined accurately. In addition, the value of pi has never been verified against the dimensions of the circle and the numeric facts have never been reconciled with the geometric facts!

The study of the numerical aspect of the circle led me to the quasi-circle theory, a modern concept of the circle that is more appropriate to the space age. The quasi-circle theory allows us to calculate the inner and outer dimensions of the circle by using unprecedented methods of calculations. Using this theory, the number of squares that makes the area of a circle and the number of arc units that makes its circumference can be computed precisely and the dimensions of the circle can be verified against its ratio.

The degree of roundness of a circle was not perceived or expressed either by mathematicians before this work, but now it can be seen with clarity. Further, it was not known that a ratio contains a certain number of individual squares that are specific to a ratio, albeit these squares are inherent to every ratio. I have called these squares "monads", a subject that will be discussed later in the study of the monad conjecture.

The quasi-circle theory is the study of approximate circles. The term "quasi-circle" is new; it is being introduced for the first time in this work to describe circular-shaped figures that closely resemble a perfect circle but are not perfect circles. In mathematical terms, quasi-circles can be defined as circles that are slightly disproportionate from their diameters. They are the newest members of the circle family. They also have their own identity and possess their own mathematical and geometrical attributes, which differentiate them from the perfect circle.

The difference between the perimeter of a quasi-circle and a perfect circle can be so minute that it may be impossible to detect with the naked eye or any known instrument; however, this difference may become highly perceptible through calculations. To better illustrate the image of quasi-circles, imagine adding or subtracting with precision a trillionth of an inch to or from the perimeter of a perfect circle while the radius stays the same. This circle is no longer perfect, but it does not fall into the category of any other known circles such as an ellipse. Thus, a new terminology was necessary to describe this new type of circles, to which I have given the name of quasi-circles.

Quasi-circles are circumscribed within a specific range of ratios. Unlike the perfect circle that uses a unique irrational number (pi) as the approximate ratio of a circle to its diameter, each quasi-circle carries its own particular ratio; and this ratio is always a rational number. To refresh your memory, an irrational number is a number that cannot be expressed as the quotient of two integers.

The author has identified two types of quasi-circles: a concentric and an eccentric quasi-circle. In mathematical terms, a concentric quasi-circle has a ratio slightly smaller than pi. An eccentric quasi-circle has a ratio slightly greater than pi. Both concentric and eccentric quasi-circles are calculated in the same manner, and the difference in their identity makes no difference for computational purposes.

Theoretically, a quasi-circle is made of a specific number of tiny individual squares tightly packed together that can be arranged in a circular pattern to create the illusion of a perfect circle. The size of these individual squares may range from minute to subatomic dimensions depending on the ratio of the quasi-circle. For this reason, this theory may be of importance to those interested in the theoretical physics of subatomic particles, or nanotechnology. There is an infinite number of quasi-circles, and each one has its own resolution. The term "resolution" in this case means the specific number and size of small squares that form the area of a quasi-circle. The geometric precision of a quasi-circle is strictly calculated according to its ratio.

The quasi-circle theory is the result of many years of research. It brings to life a missing link from the circle family, along with an alternate method to calculate the area and perimeter of circles. In addition, it brings us a more profound understanding of the common fraction and its equivalent ratio.

Previously there was only a unilateral view of ratios; a ratio simply represented the relation between two quantities expressed as the quotient of one quantity divided by the other. Now another aspect of the same ratio can be analyzed. This new theory opens new avenues that allow us an examination of the ratio itself, the quotient, from a different perspective. Neither the Greeks nor the modern mathematicians ever explored this aspect of a ratio before! In this work, a ratio (quotient) can be viewed as the symbol of a geometric representation, such as the area of a square, a rectangle, or a quasi-circle. The reader will discover that the quotient of a fraction is not just viewed as the relation of one quantity to another but as the actual integration of two quantities merged together to form a geometric representation that is proportional to the two given values. The reader, at this point, is not expected to comprehend this statement until he familiarizes himself with the monad conjecture.

The monad conjecture is another theory developed by the author that will be found in a later chapter. It is the study of squares called monads that are inherent to a ratio or quotient and how these squares relate to a closed geometric figure such as a square, a rectangle, or a quasi-circle. The monad conjecture is the backbone of the quasi-circle theory.

This new concept allows us to take a deeper look at irrational numbers and may change forever our view of pi as an irrational number. It provides new materials that permit us to apply common fractions to geometry by using unprecedented methods of calculations. Further, it gives us an additional reason to calculate pi beyond two trillion digits—hum, just a joke! The new theory makes sense; it brings some rationality within the irrationality of pi as a number and only deals with limited parts of this number. Equally important, the same rationalization that applies to the ratio of quasi-circles can also be applied to the perfect circle in a numerical sense.

The quasi-circle theory is necessary to achieve completeness in the study of the circle and will remain an essential part of the study of the circle as long as pi exists.

In this work, the author will demonstrate that the role of pi as an irrational number fits the quasi-circle theory better than the unique role of approximate ratio that it occupies in the calculation of perfect circles. Indubitably the function of pi was not fully understood by the scholars and therefore could not have been completely justified in its actual role! It will become more and more obvious to the reader after a careful review and consideration of this work that the meaning of pi has become much broader and its function much more extended than its former role; it has now become a guide in the calculation of quasi-circle fractions, even though it does not play an actual role in the calculation of quasi-circles itself, once the fraction is calculated.

The researcher will also introduce overwhelming evidence indicating that the true meaning of pi was not well understood by mathematicians before this work. When the scholars calculated pi, they thought they were just calculating the approximate ratio of a circle to its diameter; but they were in reality calculating the measurements of an infinite number of quasi-circles without having a complete understanding of their calculations. Pi, as long as it remains irrational, can only lead to the measurements of the perimeter and area of quasi-circles—never to the precise calculation of the area or perimeter of a perfect circle.

There are two major reasons reported in mathematical history explaining why pi can never lead to the precise calculation of the area or perimeter of a perfect circle. First, the famous French mathematician Johann Henrich Lambert demonstrated the irrationality of pi in 1768. Second, the proof of the transcendence of pi was given in 1882 by another renowned mathematician from Germany, Carl Louis Ferdinand von Lindemann, meaning that pi is not capable of being a root of any algebraic equation with rational coefficients. The latter proof implies that it is impossible to find a square equal in area to a given circle. The mere idea of finding squares equal in area to circles may raise a lot of eyebrows from the scholars.

Carl Louis Ferdinand von Lindemann
1852-1939
German Mathematician

Johann Heinrich Lambert
1728-1777
Swiss German Mathematician, Physicist
and Astronomer

It is not enough for us to calculate an infinite ratio or an infinite number such as pi. We must also define our own limits of this number, or our calculations in practice will lack completeness when the time of demonstration arrives! In this case, the word "define" means to determine or set down the boundaries of an infinite number that we wish to work with, using common fractions, in order to arrive at the degree of mathematical or geometrical accuracy that we wish to achieve.

In practice, we can say with certainty that no one has ever used pi to its full extent, and we can also say with certainty that our calculations of the area and the perimeter of the circle have always been approximate. It is important that we determine the number of squares that forms the area of a circle, and we must find the equivalent of its perimeter in a straight line. Further, we must be able to verify the validity of the ratio of the circle twice: first, as the proportion of the perimeter to its diameter; and second, as the proportion of the area of the circle to the square formed by its radii. These are geometric and numeric facts that we have never been able to achieve with precision.

The answer to these problems will be found in the monad conjecture and the quasi-circle-theory.

In the words of Martin Luther King, "It is better to know where we are going and do not know how, than to know how but do not know where." In the case of the calculation of pi, we knew how but not where. The scholars reckoned an infinite number through the aid of infinite series, calculus, and other approximate methods of computations; but they never fully comprehended exactly where this number had led them. Mathematicians, after two millennia of observation, analyses, and calculations failed to understand completely the results of their own calculations! The scholars failed to go the extra mile, to discover a step that was so obvious but never explored, the final step that would have led them to the quasi-circle theory. Nevertheless, to understand the final step, we must modify our old concepts of fractions, refine our mathematical techniques, and transform these into more modern concepts that adapt better to the field of geometry. This new perception of the fraction will change forever the views of the ratio in mathematics.

Mathematicians were aiming at calculating the exact ratio of the circle to its diameter, but our present methods of calculation led us to unexpected results, generating an irrational number that would inevitably bring us to the calculation of "approximate circles."

(Note: for the reader's information and for the purpose of this study, the author may substitute the term "approximate circles" to designate quasi-circles.)

According to Confucius, long ago in China, when an archer shot at his target and missed it, he turned around slowly, bowed his head, and closed his eyes to try to find the cause of his failure within himself. Perhaps we are too reluctant to admit that the reason for our own failure often lies within ourselves. Our knowledge of the perfect circle is very limited. We cannot measure with precision the length of the perimeter or the area of a perfect circle after two millennia of calculations. Along the years, we failed to pay attention to the results of our own calculations and ignored constantly the obvious clues pointing toward the existence of quasi-circles. When we calculated pi, we blindly ignored the screaming facts begging us to acknowledge the existence of approximate circles! We restricted our views to express exactly what we had set out to do in the first place, which was to calculate an approximate ratio, not the dimensions of an approximate circle. It is difficult for us to think outside the box if we are totally unaware that we are inside the box!

It is peculiar that modern scholars did not even explore the idea of quasi-circles after more than two thousand years of calculations! Indubitably when the modern scholars were calculating pi, they were aware they were confronting a very famous problem in geometry that has baffled mathematicians all over the world since the golden years of the Greeks. This problem is known as the "quadrature of the circle," or the "squaring of the circle". The problem is to find a square equal in area to a given circle. The Greeks, at least, could not solve it by the use of compass and unmarked ruler alone, and the problem remains unsolved until now!

If we take a retrospective look into mathematical history, we can divide the recorded history of pi into three distinct periods:

The first period began with the Rhynd Papyrus in Egypt in approximately 1550 BC when the Egyptian scribe Ahmes used the formula $(d - 1/9d)^2$ to calculate the area of a circle and ended just before the discovery of decimal fractions by Jamshid Al-Kashî in the early 1400s. Thus, the end of the first period in the east coincides with the beginning of the Renaissance period in the west. Al-Kashî was born in Kashan, Iran, near the Central Iranian Range. He was an assistant to the prince-astronomer Ulugh Beg and director of the observatory at Samarkand around 1424. Samarkand is one of the oldest cities in Central Asia and now the second largest city of Uzbekistan.

In the west, the first period ended with Fibonacci, who gave the value of pi as $864/274 = 3.141818$ and its limits as 3.1410 and 3.1427 in the early 1200s.

The second period was marked by the discovery of decimal fractions by Al-Kashî and ended right before the invention of the desk calculator.

The third era started with the invention of desk calculators and programmable computers, giving mathematicians access to rapid and accurate calculations. The latter is still going on today and is highlighted by the work of Fabrice Bellard, a French programmer who announced he calculated pi to 2.7 trillions digits on December 31st of 2009 by using a desktop computer. Bellard shattered the previous record set by Daisuke Takahashi by 123 billion digits and became instantly famous. We have to take our hats off to this chap! The previous records set by Takahashi of the University of Tsukuba in Japan was 2.577 trillions digits in August 2009. He used a T2K open supercomputer at the Center for Computational Sciences at the University of Tsukuba.

From the first period to the third period, the influence of the Greeks was so strong in geometry that no one considered the existence of approximate circles for over two millennia. Any problem that could not be solved by the use of compass and unmarked ruler alone was considered evil and excluded from Greek geometry. Decimal fractions were unknown to the Greeks and were not known to the west until the late 1400s. They were first introduced by al-Kashî, in his *Al-Risâli al Mohîtîje* (*Treatise on the Circumference*) written around 1424. He gave the value of pi correctly to sixteen decimal places and to a higher degree of accuracy than all his predecessors. His value was 3.1415926535898732, and he used the word *sah-hah* to describe it, meaning complete, correct, integral. At the time al-Kashî introduced his *Treatise on the Circumference*, there was no evidence to support that decimal fractions were known in Europe before 1424. It took another two hundred years before Ludolph Van Ceulen broke Al-Kashi's record by calculating pi to twenty decimal places.

Before Gutenberg invented printing in 1438 at the beginning of the Renaissance period, operations involving common fractions with large terms were infrequently found. At that time, the term "common" was used to differentiate common fractions from sexagesimal fractions found in the study of astronomy. Christoff Rudolff was the first European to submit evidence that he fully understood the significance of decimal fractions in his *Exempel-Büchlin*, published at Augsburg in 1530, a century later. He used a bar instead of a decimal point to write decimals. According to some historians, his work was not fully understood until Simon Stevin, a Flemish mathematician, published his work on the same subject in 1585. It is said that it was the translation of this work, titled *Disme, The Arts of Tenths or Decimal Aritmetike*, that inspired Thomas Jefferson to propose a decimal currency for the United States. The word "dime" which represents 1/10 of a United States dollar may have come from the title of his book, *Disme*. (For a more detailed history of pi, please consult the chapter "Pi throughout History" in this book.)

Archimedes established the limits of the ratio of the circle to its diameter to be less than 3 1/7 and greater than 3 10/71. These parameters expressed in modern decimal forms are equivalent to 3.14285714 . . . and . . . 3.14084507. Since Archimedes established these fractional limits, mathematicians have not stopped calculating pi until today, trying to find more and more accurate limits. However, it was only after the invention of decimals that the scholars made some real progress in the calculation of pi.

For centuries, mathematicians confronted continually by the problem of the quadrature of the circle opted finally for the only viable solution that was available. This solution was to search for repeating decimals in the calculation of pi, and if repeating decimals were found, they could be converted into an improper fraction and therefore into a rational number. Such a solution would bring their calculation to an end. Thus, the problem would be over finally; they would have found, not the golden ratio but the platinum ratio, the greatest prize ever dreamed of by researchers in mathematical history! However, despite of centuries of relentless efforts, instead of finding repeating decimals, frustrated mathematicians were rewarded with an irrational number that became the longest number ever calculated in human history.

The modern mathematicians never realized that when they opted for decimals instead of fractions, they inadvertently discarded a fundamental element that is essential in the calculation of circles. This brings to mind Jesus's statement found in Matthew 21:42 of the King James Version of the Bible:

"The stone which the builders rejected, the same is become the head of the corner."

This statement may come as a surprise to the scholars, but it is the perfect expression for the circumstance. Mathematicians never fully grasped the complete significance of a fraction in the calculation of circles; thus, fractions were dumped for decimals, like precious stones tossed away by the blind who is unaware of their worth. Mathematicians appeared to be completely satisfied with pi meeting all their needs for the calculation of the circle. They became strictly interested in the decimal equivalent of the ratio and lost practically all interest in the fraction itself.

In this work, I am restoring the stone that was rejected inadvertently by the builders by bringing back fractions along with their decimal equivalent. Mathematicians of all nations are urged to find better fractions that represent pi in order to arrive at better calculations of the inner dimensions of the circle instead of searching new limits for pi. At least with these fractions and using my formula, mathematicians will have a much more meaningful tool to calculate the number of squares that makes the area of a circle or compute the number of rectified arc units needed to make the circumference of a circle of a certain ratio. In addition, they would be able to verify geometrically the dimensions of the circle against its ratio.

Mathematicians have now calculated pi over two trillions and seven hundred billion digits with the aid of programmable computers, but repeating decimals were never found! For the past 150 years, the proof of pi's irrationality offered by Lambert in the seventeenth century and the proof of its transcendence given by Lindemann in the eighteenth century stood firm and did not concede any ground. There are no more cigars for calculating pi to a new limit. New records are probably being broken as this is written, but no noteworthy contribution has been made in the past two millennia regarding the methods of computations of the circle itself.

Nevertheless the efforts of countless mathematicians who contributed to the calculation of pi were not completely lost. In fact, their efforts confirmed further the proof offered by Lindemann and Lambert and laid the perfect ground to develop the quasi-circle theory. To honor and commemorate these proud and untamed mathematicians, I have dedicated a chapter to the history of pi.

Arguments in support of the quasi-circle theory

For many years, we managed to avoid successfully the obvious question that an approximate ratio would inevitably lead us to the calculation of an approximate circle. We also knew that if pi were not perfect, the calculated dimensions of the perimeter of the circle would also be inaccurate. If the perimeter were inaccurate, then we should have also known that we were no longer dealing with perfect circles, only approximate circles!

The mere fact that we call pi an approximate ratio means we knew our calculations failed to produce an exact ratio. If pi is an approximate ratio, we must automatically assume there is an *exact ratio (er)* of the perimeter of the circle to its diameter that mathematicians never found or were never able to express in finite terms.

The quasi-circle theory is the study of the circle from a numerical point of view. Pi has now been calculated over two trillion and a half digits, and a new quasi-circle is born with each new digit added to the remainder of pi.

Due to its transcendence, pi generates an infinite number of quasi-circles, and each quasi-circle has its own ratio. Each new digit added to pi, at the moment it is computed, creates a new independent ratio with a new decimal value that is specific to the dimensions of one particular quasi-circle. Thus, each new quasi-circle has dimensions more precise than its predecessors, and these dimensions approach closer and closer to those of the perfect circle, without ever reaching them, as our calculations extend further into infinity. If we could suspend suddenly our calculations in time just before the next digit is calculated, this new value of pi represents an independent ratio that could be expressed as a fraction at that particular moment in time. This fraction is a rational number that contains all the necessary elements for the calculation of quasi-circles.

According to Dr. David Eugene Smith, in his *History of Mathematics*, volume II, William Jones, an English writer, was the first to use π as the ratio of *c/d* in his *Synopsis Palmariorum Matheseos* published in 1706. In page 243, he used the word "periphery (π)" in reference to pi and was more definite in page 263 by giving 3.14159, & c. = π. Renowned Swiss mathematician Leonard Euler adopted the symbol in 1737, and it has been in general use since that time.

If we start with a fraction such as *c/d* (*circumference/diameter*) and such a fraction exists, the ratio of *c/d* is a rational expression, and the ratio of the circumference to its diameter is simply an unknown rational number that we were not able to express in finite numerical term.

c/d = er (*circumference/diameter* = exact ratio)

In this case, the exact ratio (er) is theoretically an unknown rational number, and the dimensions of a circle cannot be computed with an unknown ratio.

We have a tendency to represent pi as the approximate ratio of *c/d* (*circumference/diameter*), but irrational and transcendental numbers cannot be expressed as a fraction, and pi can never be represented as the ratio of *c/d*. The idea of pi being expressed as the result of *c/d* was deeply embedded in our subconscious mind long before English writer William Jones proposed it in early 1706. Keep in mind that when William Jones established the relation between pi and *c/d*, it was not known at the time if pi was rational, irrational, or transcendental. Lambert, who first proved the irrationality of pi, was born in 1728, twenty-two years after William Jones published his book. Further, the Fabius universal ratio formula invented by myself was not known as yet (it will be found in a later chapter), and the notion of quasi-circles was not offered until this work. So the clashing difference between the rationality of *c/d* and the irrationality of pi, even though it may have been apparent in previous centuries, became less and less apparent in the later years since pi has not been represented by a fraction recently in too many mathematics textbooks. Common fractions began to gradually disappear from the picture of the ratio of the circle since the work of al-Kashi, followed by the work of Rudolf and Stevin. Today, decimals have overshadowed common fractions almost completely and fractions are now almost inexistent in the case of pi.

The fractional expression *c/d* = pi (*circumference/diameter* = pi) is inaccurate and unacceptable, but in practice it is the formula currently being used at this time even though many scholars may try to argue or deny this fact.

In the fraction above, since pi is irrational, it makes *c* (circumference) automatically irrational. If *c* is irrational, then *c* cannot be represented by an integer. Therefore, *c/d* = pi is not an acceptable rational expression unless the definition of rational numbers is changed. So pi stands alone in the calculation of the dimensions of a circle and cannot be represented by a fraction.

Transcendence is strictly a characteristic of pi but not characteristic of the ratio of *c/d*. Lindemann's proof relates strictly to the transcendence of pi but not to the ratio of *c/d*, which by definition must be a rational number since any quotient generated from the results of *c/d* must be rational.

The author uses the following fractional expression to determine the dimensions of quasi-circles:

½C/R = *QCR; meaning (1/2 approximate circumference/Radius = quasi-circle ratio)*

In the quasi-circle theory, *c* is changed to a capital *C*, because the capital *C* automatically represents an approximation of the circumference of a circle without having to use the approximate (≈) symbol every time. It is easier to think of ½C/R as half of the expression c/d. But in the calculation of the perimeter of a quasi-circle, C represents the number of arc units that makes the circumference of a circle, and R represents the divided base of the radius (this will be explained later in the study of the monad conjecture and the quasi-circle theory). The capitalized letters in the fractional term ½C/R is used strictly in connection with quasi-circle ratios. Instead of using pi, the expression QCR is used, meaning quasi-circle ratio. A quasi-circle ratio is different from pi as it is always represented by a fraction.

A quasi-circle ratio is the quotient, or decimal equivalent of any fraction ½C/R *(1/2 approximate circumference/radius)*, that derives from the first three digits of pi, 3.14 . . . and 3.14 . . . automatically establishes the range of ratios for quasi-circles. Any fraction that the quotient starts with 3.14 . . . is called a quasi-circle fraction. Quasi-circle fractions can be calculated ad infinitum with each new digit added to pi. The criteria that define a quasi-circle ratio will be disclosed in a future chapter that focuses on the rationalization behind using 3.14 . . . as a quasi-circle ratio.

Note that ½C/R is always represented by integers, and the expression is always rational. To be clear, this fractional expression ½C/R is replaced by actual integers and contains all the elements necessary to calculate the dimensions of a quasi-circle. The author uses the word "integer" intentionally because he wants to leave open the possibility of using fractions with negative integers to represent the dimensions of the circle, just in case scientists find an application for the quasi-circle theory in physics, where particles carry positive and negative charges. Nevertheless, for the purpose of this study we will be concerned strictly with positive integers.

For example, in this work, old fractions and ratios such as 22/7 = 3.1428571 introduced by Greek mathematician and inventor Archimedes of Syracuse and 355/113 = 3.141592 given by Zu Chongzi of China are both considered quasi-circle fractions, and the decimal equivalent of these fractions are called quasi-circle ratios. In other words, a quasi-circle ratio is always generated from a fraction and always starts with 3.14 . . .

In summary, two major clues that could have led us toward the existence of quasi-circles were inadvertently ignored. The first clue was that pi is an approximate ratio, not a precise ratio; the second was that an approximate ratio leads to the calculation of the dimensions of an approximate circle, not a perfect circle!

A difference has also been established between a quasi-circle ratio and pi. Pi cannot be represented by a fraction, but a quasi-circle ratio is a number deriving from pi that always starts with 3.14 . . . and originates always from a fraction.

Do we have proof that pi is a constant?

Pi is a constant, meaning it does not vary for circles of different diameters. In other words, the proportion of the circumference to its diameter remains the same regardless of the changes made to its diameter. However, modern mathematicians greatly misunderstood the relationship of c/d and pi because they were not aware of the concept of quasi-circle ratios before this work. But once the idea of quasi-circle or approximate circumference is mathematically introduced in the study of the circle, then this notion must be considered in the hypothesis context of the circle.

The mathematical evidence of quasi-circles could have been introduced in mathematics centuries ago since the scholars accepted pi as an approximate ratio of the circle to its diameter. Unfortunately, they did not perceive before this work that pi is conducive to quasi-circles, not perfect circles. As a comparison, the fact that a planet has not been discovered does not mean the planet does not exist!

Keep in mind that pi represents physically half the circumference of a circle of unit radius and at the same time the area of a circle of unit radius; thus, the ratio must be verified against such dimensions geometrically first in order to be proved a constant. Mathematicians may prove that the relationship of c/d (*circumference/diameter*) is a constant, but this proof cannot include pi because pi is transcendental and cannot be measured accurately. The fractional expression c/d = pi is false because pi is irrational, and irrational numbers cannot be represented by a fraction.

Once more, pi stands alone. If pi stands alone, it cannot ever represent mathematically a relation between two different values, because the idea of proportion can only be expressed by a fraction that represents the magnitudes involved. This is the reason I said before that pi could only be considered as a guide in the calculation of circles.

Even though each quasi-circle has its own particular ratio, the relationship of ½C/R (approximate *circumference/radius*) remains proportional for each quasi-circle just like the perfect circle, but mathematicians perhaps did not perceive the idea found in the next paragraph before this study.

The ratio of quasi-circles is as constant as the ratio of perfect circles. For instance, if quasi-circle fractions and their decimal equivalent (44/14 = 3.1428571 and 710/226 = 3.1415929) are used to represent the circumference of a circle to its diameter, the ratios are different for each quasi-circle. Multiplying the numerator and the denominator by 2 for each fraction doubles these fractions to 88/28 = 3.1428571 and 1420/452 = 3.1415929. We find that the fractions have doubled while the ratio stays the same for each fraction respectively.

This exercise is proof that while the fraction *C/D (twice 1/2C/R)* that represents the approximate circumference of a quasi-circle to its diameter may vary for each quasi-circle; the ratio remains constant for each quasi-circle fraction respectively. Some of these explanations will become clearer as we learn to compute quasi-circles. When the fraction is doubled, the diameter and the circumference are also doubled. The ratio of a quasi-circle is as constant as the ratio of the perfect circle even though it varies with each quasi-circle individually. Pi was perceived as a constant, *but no distinction was made as yet between perfect and quasi-circles*. In reality, pi belongs to the context of quasi-circles while it does not belong to the context of quasi-circle ratios. This is why pi is used only as a guide in this work to calculate quasi-circle fractions. Pi and the old fractions offered by the Chinese and the Greeks are all considered approximate ratios of the circle to its diameter. But only the relationship of ½C/R (approximate *circumference/radius* = QCR) or *c/d* could be proved *numerically* as a constant. Pi is the ratio of the circle to its diameter, but it cannot be used to prove the relation between two given values.

It is a fallacy to think that if the circumference of the circle to its diameter *c/d* is a constant that we have also proved that pi is a constant. The definition of pi as a constant may have to be revised. The expression of *c / d* or ½*C/R* may represent the relation between two given values while pi cannot be expressed as a constant because *the idea of proportion is not involved where there is no fraction.*

Pi is conducive strictly to the existence of approximate circles, not to the existence of perfect circles. If pi is conducive only to quasi-circles from a *numerical* point of view, then the ratio of the perfect circle is unknown; if the ratio of the perfect circle is unknown, then how is pi declared to be a constant for the perfect circle! *If pi is an approximate ratio, should we also call it an "approximate constant" since irrational and transcendental numbers imply approximation?*

We must question whether mathematical logic requires exactness first in order for pi to be declared a constant, *especially a constantly changing constant! Is 3.14, 3.1415, and 3.14159 . . . , and so on the same constant? Which of these constants apply to the perfect circle?* And if exactness is required for the proof of *c/d* as a constant, then the proof of this constant can only come from the fractional expression of *c/d*, not pi, since pi cannot be represented by a fraction, and without a fraction the idea of proportion is not involved.

The *proof of pi* as a constant for the perfect circle would require the same proof as the quadrature of the circle. If pi has been proved indeed to be a constant, then the geometric facts must agree with the numeric facts. This problem has been demonstrated to be impossible with the proof of the transcendence of pi given by Lindemann in 1882. Pi cannot be considered as a constant because the circle cannot be squared geometrically with compass and unmarked ruler alone. We must thus revisit the definition of this *irrational and transcendental constant*!

The author is more inclined to accept the term *provisional constant* in the case of pi, meaning it requires further numerical and geometrical proof.

Pi relates better to the quasi-circle theory than the perfect circle approach because when pi is analyzed, it can be used as a great guide to calculate quasi-circle fractions that are in turn used to calculate approximate circles with great precision.

To reiterate, pi needs the quasi-circle theory to achieve completeness, and the quasi-circle theory needs pi for the same reason. The study of the circle is simply not complete.

Irrationality and infinity

There was always something mystical about pi. Deep down inside, there is a more profound meaning associated with this number that always puzzled us, a meaning that we all felt was within our reach but, for some vague and indeterminate reason, always eluded us and was never fully grasped!

Infinity belongs to the unknown and only expresses our own inadequacy to comprehend certain ideas of limits or limitlessness in association with certain concepts such as numbers, space, or time, etc. Here we will refer briefly to the type of infinity mostly associated with irrational numbers. This matter will be briefly covered since infinity is not the subject of this book.

Medieval writers reasoned that the number of points that makes the circumference of a circle is considered infinite, an idea that is generally accepted by modern mathematicians. Following this assertion, they were confronted with the problem that if we increase the diameter of a circle from one unit to two units, then the circumference would be equal to 2 multiplied by infinity (2 • infinity). This assertion would automatically mean that, at least, some type of infinity is measurable or could be confined to the circumference of a circle, and these concepts are more rational than irrational since the circumference of a circle itself is finite.

Pi is known as an irrational number and a transcendental number; however, when we look at a circle, we see a finite circle with a finite diameter. There is nothing irrational or transcendental about the structure of the circle itself, neither its circumference nor its diameter. It is puzzling that we came up with an infinite ratio and a transcendental number for a finite circle with a finite diameter! Thus, when we talk about the circle itself without referring to pi, there is no need to refer to its irrationality or transcendence because these traits are not characteristics of the circle itself but strictly characteristics of pi and used only in a mathematical sense.

Irrational numbers are non-terminating decimals that cannot be measured accurately and cannot be represented by two integers on the number line. To those who are interested in the philosophy of numbers, non-terminating decimals are associated with the concept of infinity because they express the idea of being boundless and unlimited.

Irrational numbers are not so irrational. In fact, they are just numbers waiting to be rationalized. Any irrational number taken to any decimal extent, or limit, is a rational number. Every decimal in pi is rational, 3.1415926535 . . . .

The difference between an irrational number and a rational number simply lies at the point where our calculation of the irrational number has stopped. If we should suspend in time the calculation of an irrational number just before the next digit is computed, then this number represents a rational number that can be converted into a fraction. If we consider the first four digits of pi, the following example will support this fact:

Pi = **3.141**5926535

*3/1=3; 31/10=3.1; 314/100= 3.14 =157/50; 3141/1000=3.141 . . .,
etc., and this exercise can be repeated ad infinitum with pi taken to any
decimal extent.*

At a glance, the numbers in the above example indicates that an irrational number can become a rational number as soon as it acquires existence since a rational number in principle is a number that can be represented by two integers. Therefore, there is a paradox between rational and irrational numbers, and it is safe to say that irrational numbers are subject to infinite rationalization!

When we talk about infinity like in the case of pi or irrational numbers, we are simply referring to their mathematical aspects as numbers that cannot be expressed as the ratio of two integers. Ironically in the case of pi, we are dealing with an infinite number that lies within certain parameters; these parameters can be established between the limits asserted by Archimedes since 225 BC, 3 10/71, and 3 1/7. In a mathematical sense, we can safely conclude that the concept of infinity associated with irrational numbers lies within certain boundaries, but we simply cannot pinpoint precisely the limits of irrational numbers within their respective boundaries.

The concept of infinity associated with irrational numbers therefore exists in a state of confined irrationality! In the case of pi, this confined irrationality will never exceed the parameters established by Archimedes. All transcendental numbers are also irrational and are also imbued of the same logic that applies to irrational numbers.

The great German philosopher Immanuel Kant expressed that our first judgment is a judgment of perception. However, perception varies with each individual; and if an object exists, its existence is factual whether it is perceived or not by an observer. If a planet exists, the existence of the planet is real whether we can perceive it or not, and it will continue to exist even we are not aware of its existence. However, the author believes only our perception of the finite increases, and only the finite exists whether it is perceived or not.

The word "infinity" is commonly used for the infinitely large or the infinitely small. In mathematics, the reason the infinity concept exists in our mind is because we know by experience, a posteriori, that natural numbers follow a certain sequence. To put it simply, if we chose a number, the number that comes before it is always smaller than the number that comes after it. Based on this logic only, we know we can create numbers at will and the same logic will always apply. We can apply the concept ad infinitum, but once a number is created, this number is finite.

In the case of irrational numbers, the numbers are not sequential; but if the numbers are assigned a place value in the decimal system, then they become rational based on the place value system because a denominator can be assigned to each number based on its place value.

For instance pi is irrational, but a denominator can be assigned to represent pi taken to any decimal extent. However, by definition, only rational numbers can have a numerator and a denominator, and if we go a little deeper and examine the following denominators in the following fractions that represent pi: 3 1/10 = 3.1; 3 14/100 = 3.14; 3 141/1000 = 3.141 etc . . . we will find that the denominator that indicates the place value of each decimal is based on the sequential order of the exponent of the number 10:

The first denominator is equal to 10^1 =10
The second is equal to 10^2 =100
The third is equal to 10^3 =1000
There is a rigid order for each denominator and each represent the place value of a decimal.

It is safe to say that once the digits of an irrational number has been calculated, it could have the same denominator as a rational number based on the place value of its digits in the decimal system! So when we think of irrational numbers, the "potential numbers" that have not been calculated as yet is what we refer to as irrational. Then, the concept of infinity associated with irrational numbers is not that far away from the concept of transfinite numbers expressed by Georg Cantor and the like. The term transfinite in this case means: going beyond or surpassing the finite. But, we can also use our imagination and go beyond the finite to assume that there are "potential denominators" waiting for each "uncalculated number" of an irrational number to convert it into a rational number. Once more, we end up with a paradox. So, irrational numbers are not so irrational, they can always be represented by two integers on the number line once they acquire existence!

Irrational numbers may be non-sequential, but ironically they have a very rigid order that cannot be changed. For instance, when Georg Vega (1756-1802) carried the value of pi to 140 decimal places, mathematicians knew and proved that only 136 were correct.

We can always go into transfinite territory by using our imagination, but the transfinite world is what is commonly called imaginary. Only the finite exist and only our perception of the finite increases, and beyond the finite is the creative power of our imagination!

For instance, when the French mathematician Fabrice Bellard set a new record on December 31st of 2009 by calculating Pi to 2.7 trillion digits, he added an extra 123 billions digits to the number. He increased our perception of the finite by 123 billions digits and therefore has set a new limit for what is finite; in reality, only our perception of the finite increased. We live in a finite world.

Using our imagination, we have the ability to create numbers based on the infinity concept, but numbers are rational once they have acquired existence. The power of our imagination to create is infinite, but once numbers are created, they are finite.

Mathematicians can no longer afford to formulate erroneous concepts on infinity without considering first the law of conservation of matter and energy already well established in the field of chemistry and physics. It is the principle that there is no variation in the total energy of the universe; albeit energy can be changed from one form to another. The law of conservation of matter and energy states that energy cannot be created or destroyed but can change its form. From the author's point of view, this law cannot function in an infinite universe but only in a finite universe! This means there is only a fixed amount of energy in the universe. Mathematicians can no longer afford to look at infinity strictly in the context of numbers; we must also consider the progress of other concepts in different branches of science and see how they would fit together with the basic concepts of infinity in mathematics. We should be careful in formulating erroneous concepts in every science because there is a great danger that it may affect future generations.

I find myself in the camp of mathematician-philosophers who say "they will believe in infinity if they are shown an infinite set right now," an expression borrowed from Rudy Rucker in his book *Infinity and the Mind*.

Quasi-circles and science

If we examine deeply the circle, we find ourselves entangled into a concept of perfect circle/quasi-circle duality similar to the wave/particle duality found in quantum mechanics. In quantum mechanics, there is no actual distinction between waves and particles. Particles may behave like waves or waves may behave like particles depending on the purpose that a physicist wants to achieve. A very convenient proposition indeed but at the same time, it is a phenomenon physicists have a hard time explaining! Since there is no actual distinction between particle and wave, theorists may choose at their own convenience the wave or the particle theory, whichever fits best their purpose or their work! In mathematics, if we examine a circle from a geometric point of view, we are dealing with a perfect circle; but if we look at a circle from a *numeric* point of view, we are dealing with a quasi-circle. Mathematicians and theorists are invited to use both, the perfect circle theory or the quasi-circle theory just like in quantum mechanics, whichever serves best their purpose.

The quasi-circle theory could become our best explanation for the wave/particle phenomenon in quantum physics, since the area of a quasi-circle is theoretically made of measurable squares that could range from small to subatomic dimensions. The waves can be described as near-perfect circles, and the particles can be described as the microscopic squares of a quasi-circle. Hopefully this theory will find some application in the study of subatomic particles or in quantum theory.

What does a quasi-circle look like? Looks are often deceiving! The moon on a bright night, viewed from a distance, appears almost like a perfect disc with a smoothly curved edge. Yet, the use of a powerful telescope will reveal that the moon does not have a smooth edge; in fact, it is made of peaks and valleys just like Earth! Such erroneous thoughts lead us automatically to the next question. Can a quasi-circle made of microscopic squares mimic the appearance of a perfect circle?

When we glance at the picture of the moon below, we must ask ourselves if we should believe the smooth curve depicted in the picture, or if we should believe the telescope that reveals the actual topography of the moon, which in reality is made of peaks and valleys just like Earth. When we look at a circle, are we actually looking at the curves of a perfect circle, or are we looking at a quasi-circle that creates the illusion of a perfect circle? Perhaps it would be interesting to find out using what I have called a mathematical scope to see if a quasi-circle, equal to pi carried to two trillion and 1/2 billion digits after the decimal point, would actually coincide with the circumference of a perfect circle of the same radius? Of course, we would have to request the service of Mr.

Takahashi who has already calculated pi over 2 trillion 500 billion digits with his supercomputer and is more acquainted with googols. Hum!

Looking at the moon from a distance, should we believe the picture of the moon showing a smooth edge, or a more powerful telescope that reveals the real topography of the moon, which in reality is made of peak and valleys just like Earth?

In the immortal words of Edna St. Vincent Millay, "Euclid alone has looked on beauty bare." Yet in the modern world, without taking anything away from the man who is deservingly called the father of geometry, we have learned that "beauty is only skin deep," and we must at times go beyond the bare facts to find the naked truth.

We live in a world where reality wears many different hats. Human cells have now revealed their genetic contents. We have learned that matter is made of mixtures, and mixtures can be separated by physical means into various compounds. Compounds in turn can be divided into elements (pure substances). We know now that the smallest unit of an element is called an atom.

We found that atoms are made of subatomic particles such as protons, electrons, and neutrons. Many more particles and antiparticles have been discovered since 1930; and they are grouped in particle physics as leptons (meaning light), mesons (meaning middle), and baryons (meaning heavy). Scientists suspect that fifteen of these particles belonging to the mesons, and baryons groups could be made of even smaller particles called quarks.

We have made more progress in particle physics in the last two hundred years than mathematicians did on the circle in the past two millenniums.

Time has changed considerably since the Greeks originally expressed their idea of atoms. A hydrogen atom, the smallest atom, measures 5×10^{-8}mm in diameter. Physicists have determined that it would take approximately twenty million hydrogen atoms to form a line as long as this dash "-" which is approximately one millimeter in length. In addition, mathematician-theorists have determined that pi computed to 39 decimal places is enough to calculate the circumference of the universe to the accuracy of the radius of a hydrogen atom.

It is questionable whether Archimedes or Euclid would be able to visualize such measurements, and it is now time to use concepts that are more appropriate to the space age to better comprehend the circle.

Imagine the circumference of a quasi-circle made of units equivalent to the length of the radius of a hydrogen atom! Now imagine the atom to be made of vast regions of space relatively to the particles that it contains, and the proton is 1,836 times larger than an electron! Suppose we use a unit the size of an electron to build the dimensions of a quasi-circle! Does anyone doubt that it is possible for a quasi-circle made of such infinitesimal dimensions to mimic the appearance of a perfect circle?

The idea is to help the reader understand that when the author speaks of the squares that make the area of quasi-circles, he could be imagining squares of atomic or subatomic proportions.

Similarities exist between quasi-circles and atoms of pure substances. An atom is the smallest particle of an element and displays all the properties of that element. Atoms appear to have been created by design and not by chance. Allow me to speculate that at the beginning, a model atom was created for one specific element. The atom contains a specific number of subatomic particles such as proton, electrons, and possibly neutrons, if any. (For the purpose of explanation, we will exclude isotopes from this picture.) Then replicas from the model atom are subsequently created from the original model to form the molecules that represent the element. Each replica of the original atom is always identical to the original model; and the number of proton, electrons, and neutrons, if any, is always the same for each atom belonging to a specific element.

When we calculate a quasi-circle, first we find a definite quasi-circle fraction with a definite quasi-circle ratio, and this ratio is particular to a specific quasi-circle of unit radius. The quasi-circle of unit radius is called a model circle. The model circle consists of a definite number of square units of specific size, called monads, and a specific number of rectified arc units that are particular to this specific ratio. Just like the atom, the internal structure or dimensions of a quasi-circle does not vary. Based on this particular model, other circles with various radii can be calculated using the same ratio.

The area of a quasi-circle contains a definite number of individual square units, and the author called these squares monads. Isn't it puzzling that the formula to calculate the maximum number of electrons that can be contained in the electron shell of an atom is expressed as $2n^2$? Why did nature use a formula such as $2n^2$ in connection with a circular orbit? Why is nature connecting squares with circles? Hopefully, the study of the quasi-circle theory will help to bring some answers to these enigmatic questions.

Physicists are already well acquainted with atomic and subatomic dimensions, and perhaps it is about time that mathematicians start using these ideas to arrive at a better understanding of the circle. Perhaps it is about time to revise our old concepts of the circle and learn to discern between the numeric aspect and the geometric aspect of a circle! Perhaps it is time to find out what are the true mathematical components that makes the structure of a circle!

Combining ancient concepts with new concepts to develop the quasi-circle theory

Liu Hui (ca. 263) was the first to get us acquainted with a Chinese method for computing the value of pi. He began with an inscribed hexagon and doubled its sides, repeating the process for each newly obtained polygon, and stated: "if we proceed until we can no more continue the process of doubling, the perimeter ultimately comes to coincide with the circle." Archimedes used a similar concept, but his method was different from Liu Hui's.

It is this old circle/polygon concept, used by Archimedes and Liu Hui, combined with a new understanding of common fractions used by Archimedes (22/7) and Zu Chongzi (355/113) as the ratio of the circle to its diameter, pieced together with the modern version of pi as an infinite number, that gave birth to the quasi-circle theory. The fact that 355/113 was the best approximation of Pi for over 900 years is noteworthy.

Neither the ancients nor the modern mathematicians were able to validate the dimensions of the circle against the ratio being used. Perhaps mathematicians no longer pay enough attention to the old facts that led us to the calculations of the modern pi.

Long ago, it was not unusual to find two different ratios to calculate the dimensions of a circle. Historians reported that the Babylonians sometimes used one ratio to calculate the circumference of the circle and another to compute its area. They had no idea that the same ratio was used to compute both the circumference and the area of the circle. Perhaps this is why the modern mathematician has a tendency to forget that when pi is used as the ratio of the circumference to its diameter, pi represents a length; and when pi is used as the ratio of a circle to the square formed by its radii, it represents an area. Therefore, when we use a ratio to compute the area of a circle, the ratio must be validated twice: it must be validated as a length, and it must be validated as an area.

For instance, if we should consider a circle of unit radius, we must validate that pi or the approximate ratio being used is in fact equal to one-half the circumference of a circle of unit radius; and furthermore, we must validate that the ratio is also equal to the area of such a circle. These are two different propositions involving the same ratio. The quasi-circle theory will meet both challenges; it will allow us to validate the dimensions of the area and the perimeter of a circle against the ratio being used.

We can obtain a lot more information about the circle using the quasi-circle theory than the perfect circle theory. Some may argue that the perfect circle is not a theory, but if we should restrict our views and our knowledge to the numeric aspect of the circle, the perfect circle becomes a theory and appears to be as unreachable as the quasi-circle theory viewed from a geometric point of view. While geometricians trust their compass, physicists have learned to trust their calculations.

Using the quasi-circle theory, we can determine with precision the number of square units that make the area of an approximate circle, and we can measure the exact dimension of these squares. We can also compute its perimeter in a straight line, and we can tell precisely the number of rectified arc units we need to build the perimeter of a circle. Additionally we can also verify the validity of the area and the perimeter of an approximate circle against its ratio.

Moreover we can also explain another aspect of a ratio; in this context, a ratio can be viewed as a group of individual squares in a random state that can assume various shapes while conserving the same area.

From a pure mathematical point of view and outside of a geometric concept, a ratio has no need for the square-unit concept. Yet, the author discovered that squares are inherent parts of ratios. In a modern sense, the author views the squares found in a ratio as virtual squares (intangible squares) and gave the name of monad to these squares. This discovery took place before the development of the quasi-circle theory. The name of monad is also used to describe the squares found in a quasi-circle. This notion was conserved in its geometric application in order to maintain uniformity between the monads found in a ratio and those found in a geometric figure. The monad concept also indicates that the square units are calculated from a ratio point of view and not from our present system of calculating areas of geometric figures.

Using a ratio in the calculation of the circle altered the prior form of logic used in the computation of geometric figures, resulting in mathematical difficulties in the calculation of circles

If we look at a perfect circle from a numeric aspect, we would have to reconcile the numeric values with the geometric facts in order to prove its existence. For example, if we start with numeric values equal to a square or a rectangle, we can prove the existence of a square or a rectangle beyond the shadow of a doubt. For instance, if we try to explain the area of a square of side 2, we must be able to explain how 2^2 is in fact equal to an area of four square units geometrically.

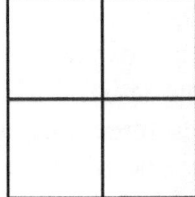

In this particular instance, we can demonstrate that the area of the square is formed of four square units, each with a side equal to 1, and we can draw precisely the area of these squares for an equivalent geometric representation. The sum of the individual squares is equal to the total area of the square, and the numeric values are in agreement with the geometric facts. This is why areas are expressed in square units, and this concept is generally accepted all over the world. We can arrive at the same conclusion using a rectangle instead of a square; but we cannot use the same logic and explain in detail, in terms of square units, the components that make the area of a circle even though the area of a circle is expressed in square units.

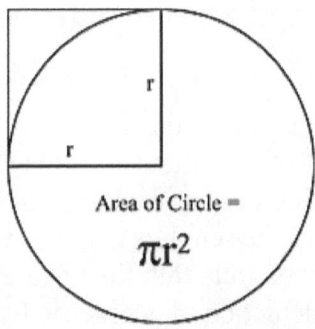

Area of Circle =
$$\pi r^2$$

The area of a circle is expressed as pi times the square of its radii (π r^2). We can determine exactly what three times the square of the radius looks like, but we cannot show with precision the geometric equivalent of 0.14159 . . . multiplied by the square of the radius; this is why pi is called an approximate ratio. Perhaps this is why we don't find too many examples illustrating the area of circles in terms of square units in many mathematics textbooks! The scholars never fully understood the dual aspect of the circle and therefore could not even try to reconstruct the circle based on the structure of its components. The dual aspect of the circle can only be understood by the study of the quasi-circle theory, and the scholars were simply not equipped with the right concepts or the right tools to reconcile the numeric facts with the geometric facts.

I submit to the scholars that this explanation of the computation of the area of circles is far from being satisfactory! The description of the area of a circle that is equivalent to three times and a fraction of the square formed by its radii may have been our best mathematical answer to compute the area of a circle, but this explanation is insufficient and inconsistent with the same logic or methodology that we have been using to calculate the area of squares and rectangles!

Just imagine that the square unit being used to explain the measurements of the area of a circle, the square of the radius, is actually bigger than one-fourth of the circle itself. The latter statement alone indicates that we never found the true components of a circle in terms of square units. The way we calculate the area of circles is not conformed to the same logic that we use to compute the area of squares and rectangles.

Did it not occur to the scholars that they were actually using a square unit that is bigger than one-fourth of the area of the circle to compute its area? There ought to be a mathematical law against this type of reasoning?

A ratio was never used before in the calculation of geometric figures! The use of a ratio indicates without a doubt that the scholars were experiencing great difficulties in calculating the dimensions of the circle. Did it ever occur to the scholars that when they introduced a ratio in their calculations of the circle that they had also altered the form of logic they used in the calculation of geometric figures? Much more than that, did they fully understand the consequences of this alteration? It is even more astonishing that we have been practicing this method for millenniums without questioning it! In a future chapter, these questions are answered fully in the study of the monad conjecture, which represents the backbone of the quasi-circle theory.

At the end of this book, the reader will realize that we never fully grasped the real concept of the circle.

We have grown accustomed to calculate the area of circles according to the only formula that was available to us, pi times the square formed by its radii; it never occurred to us that we had to revert back to fractions after we calculated pi in order to discover the true identity of the circle. Using this method, the sum of the parts can now be equal to the whole.

The quasi-circle theory gives us the opportunity to calculate the squares that makes the true component of the area of a circle, just like the square and the rectangle. This is why I believe the quasi-circle theory could reveal to be the naked truth hidden behind the bare facts! We must be careful to not let the appearance of a perfect circle drawn with a compass distort our vision from discovering another underlying identity of the circle, which could be that of an approximate circle!

Dual aspect of the circle, mathematical irregularities, and speculations on how the knowledge of the numerical aspect of the circle may affect theories on black holes in the universe

In the study of the circle, we are similarly confronted with the same problems of uncertainty that we find in quantum physics. In quantum mechanics, according to the uncertainty principle, we cannot measure with accuracy the velocity and the position of a particle at the same time. The more accurately we know the speed of a particle, the less accurately we know its position.

In this work, we are not dealing with uncertainties but with mathematical irregularities. The circle has a dual aspect, a geometric and a *numeric* aspect. Using any standard form of logic, the whole is always equal to the sum of its parts. In the case of the perfect circle, when we look at it from a *numeric* point of view, its area is expressed in terms of square units, but we cannot accurately break down its area into parts, in terms of square units, even if we fractionalize the square unit. Inversely if we start with the parts, we cannot precisely reconstruct them in terms of square units to arrive at the whole, meaning a perfect circle, as we normally do in the case of a square or a rectangle; in fact, we can only build back a quasi-circle, and the reader cannot fully visualize this statement until he familiarizes himself with the monad conjecture and the calculation of quasi-circles!

The perfect circle cannot be *numerically* subdivided into precise parts made of one standard unit with one standard value, and we do not have any numerical evidence that proves it can be built from such individual parts. In other words, the transcendence of pi implies that the perfect circle cannot be divided numerically with one standard unit having one standard value. I found this concept regarding the perfect circle to be irregular, and took the liberty to call it the "irreversibility principle of the perfect circle." Nevertheless quasi-circles can be precisely divided into component parts and therefore can be reconstructed, at least theoretically, from components parts. So if we look at the dual aspect of a perfect circle, meaning its numerical and geometrical aspects, we can never reconcile the geometric facts with the numeric values.

In summary, the circle has a dual aspect, a geometric and a numeric aspect. The Greeks and the modern mathematicians studied the geometric aspect in great details; but the numeric aspect was not fully understood before this work. If we use the geometric aspect, we cannot find the component parts of the circle in terms of square unit because the perfect circle cannot be accurately subdivided in a numerical sense. However, if we use the numeric aspect, we can find precisely the value of its component parts including the perimeter, but we cannot use these dimensions theoretically to reconstruct the area of a perfect circle but only to reconstruct theoretically a quasi-circle. The author called this the irreversibility principle of the perfect circle.

We don't possess instruments capable of drawing approximate circles; thus, the pursuit of the quasi-circle theory would appear foolish even to the most reasonable mind because it seems to point toward an obvious impossibility in geometry. The fact that quasi-circles cannot be drawn with a compass does not mean they do not exist. It only means that we are looking at the circle strictly from a geometric point of view and not from a *numeric point* of view. Thus, if we look at the circle from a numeric point of view, we will find that quasi-circles are the only circles that can exist in the universe, not perfect circles. It is very difficult to look at the circle from a numeric point of view after two thousand years of Greek geometry!

The roundness of the moon and the disc shape of the sun strongly indicate that circles built from matter of molecular size can create the illusion of a perfect circle. If we should restrict our knowledge to Euclidean geometry and the numeric world, we will find that the numeric world excludes the existence of the perfect circle and Euclidean geometry excludes the existence of quasi-circles; this is the reason why we could not count on the Greeks to develop the quasi-circle theory.

In general, celestial bodies observed throughout the universe by the means of telescopes appear to be oval and resemble quasi-circles much more than perfect circles. We have yet to find a perfect circle or a perfect sphere in the observable universe. I was starting to wonder whether examples of perfect spheres could be actually found in nature. The perfect circle appears to be an axiom, a self-evident truth, but a truth strictly found in Euclidean geometry and not confirmed by experience in nature. If examples of perfect circles could not be found in nature, perhaps the quasi-circle theory could someday supersede the perfect circle theory given enough time. Nevertheless, I decided to stay with both the perfect circle and the quasi-circle theories, relying on new observations and new theories developed by renowned scientists in recent years.

We cannot rely strictly on our powerful telescopes to assume that perfect spheres don't exist in the universe! Astonishingly enough, according to astronomer-theorists, non-rotating black holes are perfect spheres that are present in the cosmos. Hence, the black hole conjecture will be the subject of the next few paragraphs.

Theoretically, a black hole is a celestial object or dark region in space formed by the collapse of a large star with such a great mass that not even light can escape from its gravitational field; therefore, it is very difficult to detect.

Here is what Stephen Hawking, a scientist reputed as the best theoretical physicist since Einstein, had to say about black holes in his outstanding bestseller *A Brief History of Time*:

> In 1967, the study of black holes was revolutionized by Werner Israel, a Canadian scientist (who was born in Berlin, brought up in south Africa, and took his doctorate degree in Ireland). Israel showed that, according to general relativity, non-rotating black holes must be very simple; they were perfectly spherical, their size depended only on their mass, and any two such black holes with the same mass were identical. They could, in fact, be described by a particular solution of Einstein's equations that has been known since 1917, found by Karl Schwarzschild shortly after the discovery of relativity. At first many people, including Israel himself, argued that since black holes had to be perfectly spherical, a black hole could only form from the collapse of a perfectly spherical object. Any real star—which would never be *perfectly* spherical—could therefore only collapse to form a naked singularity.
>
> There was, however, a different interpretation of Israel's results, which was advocated by Roger Penrose and John Wheeler in particular. They argued that the rapid movements involved in a star's collapse would mean that the gravitational waves it gave off would make it ever more spherical, and by the time it had settled down to a stationary state, it would be precisely spherical. According to this view, any nonrotating star, however complicated its shape and internal structure, would end up after gravitational collapse as a perfectly spherical black hole, whose size would depend only on its mass. Further calculations supported this view and it soon came to be adopted generally.
>
> Israel's result dealt with the case of black holes formed from non-rotating bodies only. In 1963, Roy Kerr, a New Zealander, found a set of solutions of the equations of general relativity

that described rotating black holes. These "Kerr" black holes rotate at a constant rate, their size and shape depending only on their mass and rate of rotation. If the rotation is zero, the black hole is perfectly round and the solution is identical to the Schwarzschild solution. If the rotation is nonzero, the black hole bulges outward near its equator (just as the earth or the sun bulge due to their rotation), and the faster it rotates, the more it bulges. So, to extend Israel's result to include rotating bodies, it was conjectured that any rotating body that collapsed to form a black hole would eventually settle down to a stationary state described by the Kerr solution.

I could not thank enough the Lucasian Professor of Mathematics of Cambridge University, Dr. Stephen Hawking, a man who holds the chair once occupied by Isaac Newton, for such delightful passages and most valuable pieces of information.

Surprisingly enough, if the black hole conjecture reveals to be correct, nature would have recognized the existence of both the perfectly spherical black hole or the non-rotating black hole and the not-so-perfect black hole or quasi-sphere, also called a Kerr rotating black hole. Incredibly enough, the latest research shows that nature could be comfortable with both, the perfect circle and the quasi-circle theory.

It is amazing that Mr. Israel and Mr. Kerr working on a black hole theory came up with pretty much the same type of conclusion that I am now formulating with the quasi-circle theory. However, the celestial bodies commonly found in the observable universe seem to indicate that the universe is predominated by quasi-spheres and the proof of the existence of the perfect sphere has yet to come.

Nevertheless, based on my study of the circle, it appears that a perfect circle must be created as a perfect circle because it cannot ever be reconstructed from component parts in a numerical sense. As a parallel logic, a perfect sphere or black hole would have to be created or eventuated as a perfectly shaped uncompounded sphere. It could not be reconstructed from the elements originating from the collapse of a perfect star because the mere fact that the so-called perfect star is made of constituents is proof that the parts did not come from a perfect sphere because we don't have any numerical evidence that a perfect sphere can be reconstructed in a numerical sense, if we consider the proof of the transcendence of pi established by Lindemann in the late 1800s. Of course, one may always argue that if we take a perfect circle and break it into four quarters, we can reconstruct easily this circle from a geometric point of view, but the argument that the numeric facts will never agree with the geometric facts will always remain.

When we enter strictly in the realm of logic deprived of a ruler and compass, and all we can do is imagine a perfect circle drawn by a perfect compass and our only numerical alternative to compute the area of a circle is pi, some may be prepared to readily accept a geometric statement without being able to reconcile it with the numeric facts, or vice versa; but the author did not choose to follow that road. In contrast, the author is always very careful of mathematical statements that cannot be verified geometrically. It is this type of reasoning or logic that led him to a more profound study of the circle and the theory on approximate circles. The quasi-circle theory allows us, at least theoretically, to reconcile the numbers with the geometric facts.

If such a perfect sphere exists in the universe, it would have to be created as one giant perfect sphere, and its existence could not come from the collapse of a perfectly spherical object. It would have to represent the prototype of a giant uncompounded sphere in the universe that cannot be subdivided. This perfectly spherical black hole would also imply the existence of God since the chances of finding the existence of such a giant perfect sphere in the universe is naught to none and cannot be attributed to non-intelligence.

Nevertheless the quasi-circle theory allows us to theorize that a circle built from component parts can achieve near perfect sphericity. Similarly a quasi-perfect black hole made of constituents deriving from the collapse of a quasi-perfect star could represent a quasi-sphere that has achieved near-perfect sphericity, but this black hole could never reach perfection in a numerical sense. Nevertheless we must assume that this black hole could reach dimensions so close to perfection that it could be almost impossible to tell if it is a quasi-sphere or a perfect sphere. Such a sphere would also imply the existence of God since the chances of finding such a quasi-perfect sphere in the universe are slim and make it difficult to attribute to natural occurrences only.

In order to achieve perfect sphericity, the mass of this perfectly spherical black hole would have to be made purely of elementary particles, and the ultimate particle cannot be further divided by definition. If the ultimate particle cannot be subdivided, neither can it coalesce further to make a huge perfect sphere! Moreover, the advocates of the stationary black hole conjecture would have to provide some mathematical or physical evidence as to the size or shape of these elementary particles or how they would fit together to create a perfect sphere. Furthermore, the notion of elasticity may have to be introduced as an inherent part of these particles in order to achieve a perfect sphere. The discovery that there is a precise number of individual squares inherent to the ratio of quasi-circles may imply that the volume of a sphere made of ultimate particles will not decrease further than the volume of the particles that makes its mass. It is up to physicists to determine what that volume is and how this will affect the internal structure of a black hole. It is not likely that black holes can be allowed to collapse beyond the volume of the quantity of matter that constitute their mass. The quasi-circle theory does not support the perfect sphere theory of black holes because of the irreversibility principle inherent in perfect circles. A perfect circle or sphere cannot be built from component parts with one specific shape and one standard value. A perfectly spherical black hole can only be supported by the perfect circle approach, which does not allow for component parts.

It is a difficult proposition to emulate a theory where small squares can be arranged together in a circular manner in order to create the illusion of a perfect circle. The quasi-circle theory makes this concept theoretically possible. It makes it also possible for a quasi-sphere or a quasi-perfect black hole to exist and achieve near perfect sphericity without subdividing the elementary particles. A quasi-sphere could reach dimensions close to perfection, but it could never in a physical or mathematical sense reach perfection because of the proof of the transcendence of pi. A black hole bulging outward by its equator could fall into the category of an eccentric quasi-circle or quasi-sphere. I strongly urge physicists to use my theory to enhance or formulate their own theories.

Consequences of the introduction of the compass in the universe of numbers

Back to the quasi-circle theory, the evidence of a perfect circle became obvious mainly in Euclidean geometry where one can imagine an ideal perfect compass drawing an ideal perfect circle. The introduction of the compass in the numeric world was an odd and extraordinary happening in mathematics; it is undoubtedly a strange phenomenon in the world of numerals. Concepts deriving from the use of the compass are foreign to the numeric world, and it is not unusual that complications arose when we tried to integrate the perfect circle concept into the universe of numbers.

We found it impossible to calculate with precision the perimeter of a circle as if it was lying in a straight line, and we encountered the same problem when we tried to square the area of a circle. Mathematicians even questioned along the years if the circumference of a circle measured the same when it was curved or lying in a straight line! The author suggests books on drafting and technical drawing to find the geometric methods employed by engineers to draw the perimeter of a circle in a straight line, but these methods are not considered accurate.

The study of the quasi-circle theory will reveal that only the discovery of an exact ratio could lead to the calculation of the dimensions of a perfect circle. So far, our calculations indicate there is no such exact ratio to express the proportion between a circle's diameter and its circumference. This perfect ratio does not seem to exist in the numeric world. If we consider our present method of calculation, we use an approximate ratio to calculate the dimensions of a perfect circle—a method that is incorrect and can only lead to mathematical impossibilities when determining the perimeter or area of a perfect circle. It may also lead to possible erroneous logic in determining certain concepts such as a perfectly spherical black holes formed by the constituents of a perfectly collapsed star.

The calculations we've been using so far to determine the measurements of the perfect circle is incompatible with the appropriate form of logic we have been using. We must remember that if we deviate slightly from the truth, we are no longer part of the truth. Once more, pi is not a perfect ratio and cannot be used to determine the perimeter of a perfect circle; it can only be used in association with its rightful companion, which in this case is an approximate circle! This is why we have never been able to solve the problem of the quadrature of the circle, because to solve this problem, we need to find a perfect ratio, which cannot be found with our existing methods of calculations.

Our views were obstructed by our own methods of calculations; we took the path of least resistance and refused to question any further the results produced by our method of calculations! Certainly it has been demonstrated that it is impossible to square the circle, but it is almost intolerable to think that we allowed ourselves to calculate the ratio of a circle to its diameter ad infinitum without providing a geometric method to square the circle with each new digit added during our computation of pi. After two millenniums of research, mathematicians can only display isolated and imprecise methods of calculations that show some correlation between the geometric representation of a perfect circle and the numeric value of its ratio, pi! There is no geometric proof of the accuracy of the ratio since it cannot be measured geometrically! We don't have a geometric method to prove the accuracy of pi, and reciprocally pi cannot be used to prove the geometrical accuracy of the measurements of the circle.

When we calculated pi, we knew we were achieving greater numerical accuracy as our calculations extended into infinity, but we were not able to translate this accuracy in geometrical term. In other words, we never had the adequate means to provide a constant, reliable method of calculation to represent each level of numerical accuracy in geometrical terms.

Later the reader will realize that when the author talks about desirable degree of accuracy, he is not merely referring to the number of digits after a decimal point; but he is emphasizing on the resolution of the circle as mentioned in the second paragraph, which is to determine the quantity and size of small squares that forms the area of a quasi-circle of a given ratio or the calculation of the degree of accuracy of the perimeter of a quasi-circle as its circumference lies in a straight line.

The area and perimeters of quasi-circles can be calculated without making the use of pi; but the author suggests the use of pi as a guide in order to achieve greater accuracy in calculating the area and perimeter of circles.

The quasi-circle theory will revolutionize forever the way we have perceived and calculated circles. Pi needs the quasi-circle theory in order to achieve completeness, and the quasi-circle theory needs pi to reach fulfillment; they complement each other and are inseparable twins.

Pi throughout History

Pi from 2000 BC to the AD first millennium

The first attempts to calculate the area of a circle began long before the Greek civilization. These calculations were empirical, of course, and resulted in rude approximations of pi.

The first evidence of an approximate value of pi was found in the Rhind Papyrus around ca. 2000 BC. The Egyptian scribe Ahmes used a formula that can be expressed in modern symbols as $A = (d-1/9d)^2$ to find the area of a circle of diameter, d. That would make the value of pi equivalent to 3.1605, a value based on experiment. This formula can also be considered as the first attempt to square the circle. The Rhind Papyrus is named after a Scottish antiquarian, Alexander Henry Rhind, who bought the papyrus in 1858 in the town of Luxor in Egypt, which is situated in the proximity of the Nile, near the ruins of ancient Thebes.

Historians believe that the number 3 is probably a much older value of pi even though they couldn't find much literature to prove this fact. They indicated that such a value could be found in early Chinese works, the Talmud, and the Bible. To exemplify, the scholars often referred to the following biblical verse to substantiate their claim: "And he made a molten sea, ten cubits from the one brim to the other: it was round all about, and his height was five cubits: and a line of thirty cubits did compass it about" (1 Kings 7:23; 2 Chronicles 4:2).

The scholars found the biblical value of pi by dividing thirty cubits by ten to obtain a value of three. Some have used this argument to show that not everything in the Bible is true. We must remember that the Bible is also an accurate history book written by very wise people chosen by God and who faithfully recorded the facts of their time. If we consider the fact that the famous German mathematician Carl Louis Ferdinand von Lindemann demonstrated the transcendence of pi in 1882 and if we base ourselves on the fact that all transcendental numbers are also irrational, we must deduct that pi will remain forever an approximate number. Thus, 3, 3.14, or 3.14159 are all approximations of pi at different points in time; and they all represent a quasi-circle ratio, depending on the degree of desired accuracy one wants to achieve. The number 3 simply represents one of the oldest recorded approximations of pi. There is no more cigars for finding a more accurate approximation of pi, since Fabrice Bellard of France calculated it to over two trillion and seven hundred billion digits on the last day of December of 2009! The author points out that the number 3 could have easily been included in determining the range of quasi-circles, but the range was narrowed down to 3.14 because it appeared to be more realistic for the space age. (Consult the chapter on the calculation of quasi-circle.)

The ancients appeared to be unaware that the ratio of the area of a circle to the square formed by its radii was the same as the ratio of its circumference to its diameter. It was not uncommon to see in early Babylonian texts two different values of pi, one to calculate the area and another to compute its circumference. In India, the ratio of the area found in the *Sulba Sutras* is given as 3.088 while the ratio for the circumference is given as 3.2.

Archimedes of Syracuse (ca. 225 BC), one of the most brilliant mathematicians of his era, made the next noteworthy contribution to the theoretical value of pi. He asserted that the ratio of the area of a circle to the square of its diameter is 11:14, and the ratio of the circumference of a circle to its diameter is less than 3 1/7 and greater than 3 10/71. Pi is still known as Archimedes's constant.

If we express these limits in modern decimal form, a fair equivalent of this expression would be: $3.14285714 > \pi > 3.14084507$. Archimedes used inscribed and circumscribed polygons to determine the approximate area of a circle. It is reported that he started with a hexagon and worked his way up to a polygon of 96 sides and determined that the area of a circle lies within these results.

Many mathematicians adopted the value of 3 1/7 given by Archimedes; it appeared that it was recognized as a satisfactory value of pi during this period and was subsequently found in the works of Heron (approximately ca. AD 50?) and subsequently reappeared in the thirteenth and fourteenth centuries in the work of Dominicus Parisiensis (AD 1378), Albert of Saxony (AD 1365), Nicholas of Cusa (AD 1450), and others.

In 20 BC, the Roman Vitruvius spoke of the circumference of a wheel of a diameter of 4 feet to be 12 ½ feet. Historians deducted that he used 3 ⅛ for pi.

Ptolemy (ca. AD 150) may have taken the limits asserted by Archimedes and expressed them in sexagesimal numbers; he gave 3 8' 30" as the approximate value of pi which translates into pi equals 3.1416.

The Hindu mathematicians used different values of pi and did not appear to have adopted one definite value. Āryabatha (ca. AD 510) gave the value of pi as 3.1416 (62832/2000). His method was the following:

"Add 4 to 100, multiply by 8, add 62000, and you have for the diameter of two *ayutâs* the approximate value of the circumference." The word *ayutâs* means myriads or ten thousands.

There is a great possibility that this value may have been found by Āryabatha the younger. At any rate, the Hindu value seems to be due to the Āryabathas even though it may have been obtained from the Alexandrian scholars, whose work may have well reached India around that time.

Brahmagupta (ca. AD 628) gave a value of 3.1622, a value equivalent to the square root of 10.

The Chinese mathematicians found various values of pi, but their methods are unknown. Ch'ang Höng (ca. AD 125) gave the value of √10 and the value of 142/55 was given by Wang fan (ca. AD 265), a decimal equivalent of 3.15555.

Liu Hui (ca. AD 263) was the first to get us acquainted with a Chinese method for computing the value of pi. He expressed the value of pi as $3.141024 < \pi < 3.142074$. He began with an inscribed hexagon and doubled its sides, repeating the process for each newly obtained polygon and asserted, "if we proceed until we can no more continue the process of doubling, the perimeter ultimately comes to coincide with the circle." The method used by Liu Hui was not the same as the one used by Archimedes.

The value 355/113 = 3.14159 . . . is found in the works of Zu Chongzi (ca. AD 470). It was considered the best approximation of the value of pi for approximately 900 years; a value that is attributed to Liu Hui by some historians.

Among the Arab writers, Al-Kwarizmi (ca. 800), the scholar from the House of Wisdom in Baghdad, often called the father of algebra, gave the value of pi as 3.1416. Many historians claim that the word "algebra" is due to his most important work, an algebra treatise called *Hisab al-jabr w'al-muqabala*. The word *al-jabr* means completion and the word *al-muqala* means balancing.

To conclude this chapter, I would like to express my deepest gratitude to Dr. David Eugene Smith for preserving such valuable pieces of history in his monumental work, *History of Mathematics*, volume II, which was my principal source of information for this chapter.

Pi from the beginning of the second millennium to the precomputer era

Various mathematicians contributed to the calculation of pi from the beginning of the second millennium to the precomputer era. Years after years, they improved on their mathematical techniques and were indefatigable in their attempt to arrive at better approximations of pi.

Here is a brief summary of their best approximations for that period:

Franco of Liege (ca. AD 1066) gave pi = 3.142857, or 22/7, which was an old value already given by Archimedes as a limit.

Fibonacci (ca. AD 1220) found pi to three decimal places. He gave pi = 3.141818 and offered the limits of 3.1427 and 3.1410.

Madhava (ca. AD 1400) gave pi to eleven decimal places, 3.14159265359.

Jamshid Al-kashi (ca. AD 1430) found pi to fourteen decimal places. He used Arabic characters to express it in the following manner:

Sah-hah (meaning complete, correct)

3.14159265358979

Viète (or Vieta) (ca. AD 1593) gave the following: 3.1415926535 < π <3.1415926537

Ludolf Van Ceulen (ca. AD 1596) gave pi to thirty-five decimal places, and it is not uncommon to find German textbooks referring to pi as the *Ludolphische Zahl*.

Aedriaen Van Roomen (ca. AD 1561-1615) computed pi to seventeen decimal places, and Newton (ca. 1665) gave pi to sixteen decimal places.

Abraham Sharp (ca. AD 1717) obtained the value of pi to seventy-two decimal places.

From the early 1700s to the mid 1800s, mathematicians looking for fame tried relentlessly to find new limits for the value of pi. Among those who distinguished themselves for their outstanding work is Georg Vega (1794), who calculated pi to 140 places (136 correct), Zacharias Dase (1824-1861), who carried pi to 200 decimal places, William Rutherford (1853), who found it to 208 places but only 152 were correct, and last but not least, William Shanks (1853), a British amateur mathematician who calculated pi to 707 decimal places by using John Machin's formula, of which 527 were correct. The mathematician D.F. Ferguson found the error in 1944 using a mechanical desk calculator.

Computer era of approximation of pi

The next generation of calculation was marked by the introduction of electronic calculators and computers to compute the approximate value of pi. The new era started with Ferguson, who gave pi to 710 places using a desk calculator. The latest approximations available in

August 2009 were those of Daisuke Takahashi, who used a T2K open supercomputer at the University of Tsukuba in Japan to calculate pi beyond 2.577 trillion digits, and this record was recently broken by the work of Fabrice Bellard on December 31st of 2009, who claimed he calculated the value of pi to the incredible number of 2.7 trillion decimal digits using a desktop computer that cost less than $3000.

You have to take your hat off to these chaps!

While I find the work of Mr. Bellard and Takahashi remarkable, I realize that they are not aware of my latest work and formula $1/2C/R = (1/R)^2$ $(1/2C \cdot R)$, that would allow them to calculate the inner dimensions of the circle by using any fractions equal to pi taken to any decimal digits extent. I urge this new generation of mathematicians to use their skills to find better quasi-circle fractions which would represent quasi-circles with more accuracy than the last fraction I have provided, 105414357/33554432 = 3.14159265 . . . Another great fraction found on *Wikipedia*, the free online encyclopedia, is 103993/33102=3.14159265301 . . . Oops! I hope I have not opened a new can of worms!

In conclusion, the reason for writing this chapter is not just to commemorate the great mathematicians who contributed to the evolution of pi as a number; it is also to show beyond the shadow of a doubt that the modern mathematician had abandon any hope of calculating the inner and outer dimensions of the circle and his only hope was to set new records for pi or look for repeating decimals in the number. The concept of quasi-circles was never contemplated in the history of mathematics before this work, and the computations of the inner or outer dimensions of the circle were not fully understood before this study!

The Monad Conjecture
and
Fundamentals of the Quasi-Circle Theory

Introduction to the monad conjecture

In this study, the author will explore a new mathematical aspect of ratios as promised in the first chapter and will also seize this opportunity to offer a mathematical definition of the monad, an intangible element that represents the building block of quotient/ratios and is subsequently used to calculate the area of quasi-circles. The reader will also find out how a ratio can be used to determine the shape of a geometric figure. This new understanding of ratios is crucial to the computation of the area and perimeter of quasi-circles.

The reader may ask why is it so important to learn another method to calculate the area of geometric figures?

The answer to this question is simple; the perimeter and the area of the circle have never been calculated with precision, only approximately. Mathematicians were unable to calculate the area of circles in the same manner they calculate the area of squares or rectangles. To compensate for their inability to calculate precisely the area of a circle, they computed an approximate ratio that allows them to find its approximate area in proportion to the square formed by its radii. When they used a ratio in their computations, they departed from the conventional method of calculations that they would normally use to compute the area of a closed geometric figure. Furthermore, when they used a ratio, they also altered the form of logic used in the computations of closed geometric figures without realizing it. In other words, they changed the rules of the game in the middle of the game, and this is not allowed!

For instance, we use the dimensions of the sides of a square or a rectangle to find its perimeter or its area; for a triangle, we use the base and the height; but for a circle, we use a ratio. This altered form of logic is reflected in our methods of computations and was handed over to us from generation to generation. We grew accustomed to this method of reasoning without understanding fully its implications. It is needless to say that we accepted the scholars' method of computations for the circle without questions or investigation.

Without the proper form of logic, we would never be able to reconcile the numerical aspect of the ratio with the geometric facts. This altered form of logic takes roots with Archimedes when he inscribed and circumscribed polygons to establish the limits of the value of pi between $3 + 10/71$ and $3 + 1/7$. He used an arithmetic method instead of a geometric method to obtain faster calculations of the dimensions of the circle without using a geometric procedure. As our mathematical techniques improved along the years to make this process more accurate, the original fractions that led them to the first measurements were discarded in the process because the scholars never understood their true importance. The author discovered the importance of the fraction that was discarded by the scholars and is now trying to reinstate the latter in the calculation of the circle, but please realize that the importance of pi cannot be diminished, and the author uses it as a guide to calculate quasi-circle fractions.

The calculation of geometric figures making the use of ratios falls under the study of the monad conjecture, which is being introduced for the first time in this work. It never occurred to us that when we changed our form of logic, it would be necessary to find a way to retrace our steps back to our old form of logic in order to reconcile the dimensions of a geometric figure with the numeric facts. The scholars were simply not aware of the monad conjecture that allows us to compute the dimensions of certain geometric figure according to their ratio. They were not aware either of the quasi-circle theory and therefore were unable to calculate the exact number of squares for the ratio being used. Furthermore, they were never able to verify the area or the perimeter of a circle against its ratio.

Overview of this work's organization

In the following paragraphs, the author will get the reader acquainted first with the aim of the monad conjecture and some of the key terms associated with it.

Following this introduction, the study is organized into twelve different objectives:

The first objective will help the reader to differentiate between a monad and a square unit found in a closed geometric figure.

Objectives 2 to 6 will familiarize the reader with the *Fabius universal ratio formula* invented by the author. With this formula, the reader will also learn how to calculate the number of monads in a ratio and how to compute their individual dimensions. Furthermore, he will learn how to verify these dimensions against their respective ratios from a pure mathematic point of view, a concept that is foreign to today's mathematician.

In objective 7, the reader will see how the shapes of three basic geometric figures can be determined according to their ratio and furthermore, how their area can be calculated according to their ratio. The reader will start noticing also the gradual integration of the quasi-circle theory into the monad conjecture.

Objectives 8 and 9 will introduce the reader to the idea of model squares, model rectangles, and model circles and the calculation of model squares and rectangles. The calculation of model circles will take place in objective 10.

In objective 10, the reader will be introduced to the basics of the quasi-circle theory and the definition of key terms used in the calculation of quasi-circles. He will see also how the fundamentals of the monad conjecture learned from objectives 2 to 6 are adapted and applied to the quasi-circle theory.

The quasi-circle theory alone contains eighteen objectives and represents an integral part of the monad conjecture. Among these objectives, the reader will learn how to determine the number of monads that makes the area of quasi-circles and the number of rectified arc units needed to make the circumference of a circle for a particular ratio. He will also learn how to verify the dimensions of a circle against its ratio. These are things that we were unable to achieve before this work. Finally, objectives from the monad conjecture are listed in numerical order, and those of the quasi-circle theory are listed in alphabetical order for the purpose of differentiating two different works that are intertwined with each other but can be considered altogether as entirely separate studies.

In objective 11 the author will help the reader identify that all the previous objectives were met, and in objective 12, the author will proceed with the study of the monad conjecture to show how a rectangle and a quasi-circle of the same ratio, have also the same area, and also how the perimeter of such a rectangle could transform into an equivalent circle. Please realize that because the study of quasi-circles is so new, a separate chapter is needed for the study of quasi-circles only, but the acquisition of this knowledge is important before we can conclude this study.

What is the monad conjecture?

The monad conjecture is the study of intangible squares called monads that are inherent to a quotient or ratio; it is also the study of how a ratio/quotient relates to a geometric figure. The word "monad" in this case means "a unit; something simple and indivisible," according to the *Webster's New World College Dictionary*.

A ratio resulting from a fraction $a/b = q$ contains a specific number of individual squares (monads) that are particular to that ratio or quotient. These individual squares (monads) have specific measurements, and the sum of their areas is equal to the ratio or quotient resulting from the fraction.

Monads can be distinguished from other squares by the direction of their magnitude. Compared with other squares, the magnitude of monads moves toward the infinitely small and never toward the infinitely large. The area of a monad is always smaller than unity. In the fraction $a/b = q$, the side of a monad is found by dividing $1/b$ and its area as $(1/b)^2$.

A monad is always fractional; thus, in the fraction $a/b = q$, b is always greater than 1 ($b > 1$).

The monads (individual squares inherent to a ratio) cannot be further divided by definition, and each monad represents the ultimate intangible square for a particular fraction and its accompanying ratio. When monads are seen from a geometric point of view, they represent the ultimate square units that make a magnitude. All the monads found in a ratio/quotient are identical.

In this context, monads could easily be seen as the building blocks of magnitudes.

(Please note that the dimensions of the individual squares are not calculated from the ratio itself but from the fraction because the fraction is subject to change while the ratio stays the same.)

Objectives of the monad conjecture:

1. *Discuss the rationalization behind the monad concept*
2. *Calculate the side of an individual monad in a quotient/ratio*
3. *Calculate the area of an individual monad in a ratio*
4. *Calculate the total number of monads in a ratio*
5. *Calculate the total area of monads in a ratio*
6. *Calculate and verify that the total number of monads found in a ratio multiplied by the area of each individual monad is equal to the ratio*
7. *Identify the shape of a geometric figure based on its ratio*
8. *Introduce Model squares, model rectangles, and model circles and the relativity of measurement's values based on a value equal to unity*
9. *Calculate the area of a model square and a model rectangle according to their ratio based on the concepts of the monad conjecture*

10. *Explore the basic mathematical concepts behind the quasi-circle theory and calculate the inner and outer dimensions of quasi-circles (subdivided into eighteen objectives)*
11. *Establish that the same formula can be used to calculate the area of three basic geometric figures according to their ratio: the square, the rectangle, and a quasi-circle.*
12. *Establish that a quasi-circle and a rectangle of the same ratio are also equal in area (a proposition crucial to the quasi-circle theory)*

Objective 1. Discuss the rationalization behind the monad concept

The author, sometime in the mid-1990s, came to realize that there was a need to differentiate between a square unit originating from a quotient/ratio and a square unit originating from the area of a closed geometric figure. If we consider a ratio from a pure mathematic point of view outside of a geometric concept, the ratio has no need for the square unit concept. Nevertheless the author found that squares are inherent part of ratios. The author visualized the squares inherent to a quotient/ratio as intangible squares and the squares pertaining to the area of geometric figures as real squares. For this important reason, a new terminology was necessary to describe the intangible squares found in a quotient/ratio, and the author finally settled for the name of "monad" for these squares because their application in the study of ratios was given more importance than their application in the study of geometry. The concept of the monad led the author to the development of the monad conjecture, which represents the backbone of the quasi-circle theory. So this work will expose the reader not only to the monad conjecture, but also to the quasi-circle theory, which is considered as an integral part of the study of the monad conjecture.

The definition of monads associated with ratios and quotient of fractions took years to develop, and these concepts are offered for the reader's review and consideration as an addendum to our previous concept of ratios.

The Greeks introduced the word "monad" during their study of points and lines in geometry. Dr. David Eugene Smith, in his *History of Mathematics*, volume II, stated that Pythagoreans defined a point as a monad having position, a proposition that was later adopted by Aristotle in 340 BC.

Monad comes from the Greek word *monas* meaning unit and the word *monos* meaning alone. Hyppolytus, one of the most prolific writers of the early church, stated that the Pythagoreans called the first thing that came into existence a monad. According to Diogene Laertius, the monad begat the dyad, then from the dyad came the numbers, the numbers begat the point, then the line, then came the two-dimensional and three-dimensional entities and bodies, followed by the four elements, earth, fire, water, and air, which the world is made of, etc. In the philosophy of Gottfried Leibniz (1646-1716), a monad is an indivisible, impenetrable unit of substance viewed as the basic constituent element of physical reality.

The concept of the monad appeared to have strong support among the ancients but never achieved full maturity in a mathematical sense. The idea eluded mathematicians for centuries until the scholars finally abandoned it. Leibniz, the great mathematician and philosopher who shares the invention of calculus with Newton, revived the concept of the monad in the early 1700s, but this concept remained strictly a part of his philosophy or discussion on metaphysics but was never integrated into his mathematical works.

The word "monad," as defined in the monad conjecture, means simply a single and indivisible square found in a ratio. In the monad conjecture, there is no such thing as an ultimate square that makes all quotient or ratios; however, there is an ultimate square for each particular ratio. The essence of the monad's definition as expressed in the monad conjecture originates from Greek philosophy and Leibniz's philosophy as well, even though it does not relate at all to metaphysics in this context.

The Greeks may have considered a point as a monad having position; however, neither the Greeks nor Leibniz ever offered a mathematical proof of the existence of the monad, because, in order to acquire existence, the monad of the Greeks would have to be attributed a size, shape, and dimensions which would automatically differentiate it from the modern definition of a point. The standard and modern definition of a point being, according to *Webster's College Dictionary*, an element in geometry having definite position, but no size, shape, or extension.

Any attempt by the scholars to define the monad as a physical element would face the same theoretical difficulties as the monad of the Greeks, meaning a physical point would be automatically associated with the idea of dimensions; and mathematicians would have a real feast demonstrating that a physical point could be divided indefinitely into many parts, and therefore, it could not be the ultimate element that makes the fabric of all magnitudes.

A mathematical or geometrical description of the monad was necessary in order for it to acquire existence; after all, to describe the monad was the same as describing the ultimate particle that makes the building block of the universe! It is needless to say that the geometric concept of the monad died gradually. Today's mathematicians skipped the monad's concept, but they failed to explain how they could arrive logically at building a magnitude without this concept.

In the monad conjecture, the monad needs no further division to satisfy the needs of a particular ratio/quotient and by definition is indivisible. In the quasi-circle theory, the monad concept satisfies the most demanding concepts of mathematics even those related to the perimeter and area of circles.

To pay homage to the Greeks and Leibniz, the monad represents the ultimate element of physical reality for a particular ratio/quotient. When I discovered that each ratio contains a certain number of identical squares inherent to the ratio, I thought the monad concept would fit it like a glove.

It is now time to share with the reader the mathematical aspect of monads found in quotient/ratios since he is now a bit acquainted with the concept. In the next paragraph, the reader is invited to join the author to see how objectives 2 to 6 can be determined by using the Fabius universal ratio formula developed by the author. Later the author will progress gradually to help the reader understand the application of the universal ratio formula in geometry.

Objectives 2 to 6 pertain to the monad conjecture and are considered as a group

For the purpose of clarification, let's consider objectives 2 to 6 as a group since these objectives deal strictly with calculations pertaining to the monad conjecture:

Objective 2. Calculate the side of an individual monad in a quotient/ ratio

Objective 3. Calculate the area of an individual monad in a ratio

Objective 4. Calculate the total number of monads in a ratio

Objective 5. Calculate the total area of monads in a ratio

Objective 6. Calculate and verify that the total number of monads found in a ratio multiplied by the area of each individual monad is equal to the ratio

In order to demonstrate the efficiency of the *Fabius Universal Ratio Formula*, let's use this very famous fraction, *355/113 =3.14159203539*, found in the work of Zu Chongzi who lived around AD 470. The Chinese used this fraction long ago to represent the approximate value of pi and it remained the best approximation of pi for the next nine hundred years. It is still widely in use today and can be commonly found in many modern geometric textbooks.

Should we consider the aforementioned fraction, what is the dimension of a single monad in this ratio, and what is the dimension of its area? How many monads (individual squares) can be found in this ratio? How can we verify that the total number of individual squares are indeed equal the ratio 3.14159203539 . . . ?

Fabius Universal Ratio Formula to find the numbers of squares in a ratio:

$$a/b = (1/b)^2 (a \cdot b)$$

Let's represent a/b by 355/113 and replace 3.14159203539 by $(1/b)^2$ $(a \cdot b)$ in the following universal ratio formula invented by the author to solve this problem. This formula integrates many smaller formulae and is broken down into individual formula for explanation purposes:

$a/b = (1/b)^2 (a \cdot b)$

$355/113 = (1/b)^2 (a \cdot b)$

As the reader will see, it could be necessary to use a calculator or computer that can handle twenty decimal digits or more in order to arrive at the desired decimal accuracy.

Example No. 1:

Objective 2. Formula to determine the side of each individual monad in a ratio is $1/b$

$1/b$ = side of individual monad, or
$1/113 = 0.00884955752212389380\ldots$

Objective 3. Formula to determine the area of each monad in a ratio is $(1/b)^2$

$(1/b)^2$ = area of individual monad
$(0.00884955752212389380)^2 = 0.0000783146683373795\ldots$

Objective 4. Formula to find the total number of monads in a ratio is $a \cdot b$

$a \cdot b$ = number of monad (individual squares) in a ratio
$355 \cdot 113 = 40115$

Objective 5. Formula to determine the total area of monads in a ratio is $(1/b)^2 (a \cdot b)$

$(1/b)^2 (a \cdot b) = 0.0000783146683373795\ldots \cdot 40115 = 3.14159203539\ldots;$

The number of individual squares multiplied by the area of each individual monad is equal to the ratio.

The Fabius universal ratio formula at a glance is:

$a/b = (1/b)^2 (a \cdot b)$

$355/113 = (1/113)^2 (355 \cdot 113)$

$355/113 = (0.00884955752212389380)^2 \, 40115$

$355/113 = 0.00007831466833737959 \cdot 40115 = 3.1415929203539$

Objective 6. Consequently, we can formulate the equation to verify that the number of squares (monads) in a ratio multiplied by the area of each individual squares (monads) is indeed equal to $a/b = q$

$a/b = (1/b)^2 (a \cdot b)$; therefore $q = (1/b)^2 (a \cdot b)$

We have proved that the total number of individual squares (monads) in this ratio multiplied by the area of each individual monad is equal to the ratio.

We have never been able to verify the ratio of the area of a circle or its perimeter, but as we refine our mathematical techniques and we understand better the structure of a ratio thus, we can also understand better the internal structure of the circle. Later the reader will see how the monad conjecture concept was adapted to calculate the area and the perimeter of the circle.

Example No. 2:

Let's see if the *Fabius universal ratio formula* will work for another fraction $22/7 = 3.1428571$. Archimedes, who lived around 287-212 BC, gave this fraction to represent pi. Pi is also known as Archimedes's constant:

Fabius universal ratio formula

$a/b = (1/b)^2 (a \cdot b)$

Let's replace a/b by 22/7 in the formula to see the formula at a glance:

$22/7 = (1/b)^2 (a \cdot b)$

$22/7 = (1/7)^2 \ (22 \cdot 7)$

$22/7 = (0.142857142)^2 \ (154)$

$22/7 = 0.020408163 \cdot 154 = 3.1428571$

In the formula above:

Side of monad: $1/7 = 0.142857142$

Area of monad: $(1/7)^2 = 0.020408163$

Number of monads: $a \cdot b = 7 \cdot 22 = 154$

The area of each individual monad multiplied by the number of monads is equal to the ratio:

$(1/b)^2 \ (a \cdot b) = 0.020408163 \cdot 154 = 3.1428571$

The sum of the area of the individual monads is equal to the ratio.

This problem has now been solved from a pure mathematics point of view, and as promised, I will show its application in relation to geometric figures. As I mentioned before, this same formula can be used in the calculation of the area of quasi-circles and will be discussed further in the chapter "Calculation of Quasi-Circles." After the reader is introduced to the calculation of quasi-circles, we will integrate the quasi-circle theory into the monad conjecture to continue our study of the circle.

Please note that at this point, we have learned to determine the number of squares in a ratio and the area of the individual monads, but we have not yet learned to differentiate the geometric shapes according to their ratio. These notions will be offered in objective 7.

Objective 7. Determining the shape of a geometric figure according to its ratio and integrating gradually the quasi-circle's concept into the monad conjecture

The author has determined so far three basic types of ratios, and each can assume a specific geometric figure such as a square, a rectangle, or a quasi-circle.

It is easy to determine a geometric figure relatively to its ratio. We are fortunate that only one formula is used to calculate the number of squares in the ratio of all three geometric figures, and we are even more fortunate that the only variable to help us determine their geometric shape is the ratio itself. The ratio not only allows us to classify each geometric shape, it also helps us to determine the position of a rectangle or a square relatively to its base. The individual squares (monads) representing the area of these shapes can be verified against their ratio, a proposition that is very important in the case of a circle or a quasi-circle since it would require the solution to the problem of the quadrature of the circle to justify this hypothesis. This solution still remains impossible in the case of the perfect circle, but is very possible, at least theoretically, in the case of approximate circles. (Please refer to the chapter of "Calculation of Quasi-Circles" since we have never been able to verify the ratio for circles.)

First, let's complete our journey through the monad conjecture. In Euclidean geometry, we start first with a geometric figure, and then we calculate its area or perimeter. In this case, we are in the presence of squares found in a ratio, and we have to determine a geometric figure based on the ratio.

At this point, using the *Fabius Universal Ratio Formula*, we have demonstrated that we are able to determine the dimensions of each individual monad in a quotient or ratio. When we consider a fraction $a/b = q$, we proved that the number of individual squares (monad) in a ratio is equal to q, but we have no idea what shape these monads represent as a group; for all we know, they could assume any geometric shape, and there are no existing mathematical criteria to help us determine their specific shape.

Below the reader will find the criterion that will help to determine the three basic shapes according to their ratio.

From a numerical point of view, the individual monads found in a quotient/ratio grouped together do not represent any particular shape.

The individual squares are considered in a random position, but when the sum of their area is equal to 1, they can be shaped as a square. In other words, there is at least one instance where we can demonstrate that these individual squares (monads) can be arranged in the shape of a square if the ratio of the sides of the square is equal to 1.

When we analyze the fraction $a/b = q$:

1) If the ratio q is equal to 1, it is always symbolic of a square. The perpendicular sides of a square can always be seen as the base and height of a square, and the side at the bottom of the square always represents the base. The denominator of the above fraction, b, always represents the base of a square, and the numerator of the fraction, a, always represent the height of the square. Dividing side a by side b finds the ratio of the sides of the square, which is always equal to 1. (Please note that the notion of the base of the square being represented by the denominator of the fraction b may not seem important in the case of the square, but it is very important in the case of the rectangle.)

Choosing the base and the height of the square to represent its sides implies a ninety-degree angle, which is not expressed by s^2 or side multiplied by side.

Imagine finding four tiles, representing one square unit each, in our home, lying on top of each other. We cannot say for sure that these four tiles were originally shaped as a square of side two. As far as we know, these square tiles could assume any random position.

Various shapes could be made out of these squares, and we can only speculate that, at least in one instance, these square units could represent a shape equal to 2^2. When we are dealing with the monad conjecture, we are in the presence of squares found in a ratio, and we can only assume that at least in one instance, these square units can be arranged into one specific shape.

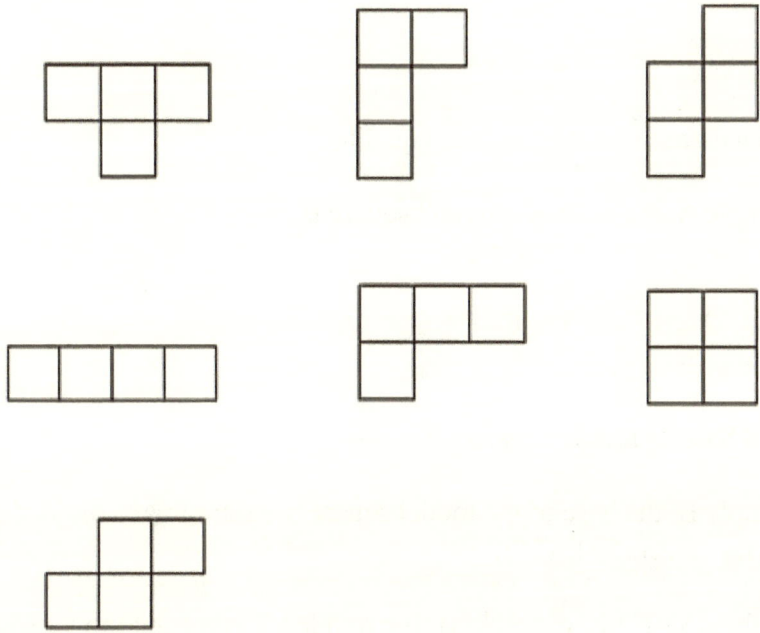

Objective 8. Introducing Model squares, model rectangles, and model circles, and the relativity of measurement's values based on a value equal to unity

In the monad conjecture, a square whose base is equal to unity is called a model square. A rectangle whose base is equal to unity is called a model rectangle. A circle whose radius is equal to unity is a model circle, and the radius is considered as the base of a circle.

Only the base of a model square, a model rectangle, or a model circle can be divided into parts by using the fraction $1/b$, to find their internal dimensions. Remember that $1/b$ is equal to the side of the monad and $(1/b)^2$ is equal to the area of a monad. The denominator of the fraction b is the dividing factor of the base and also represents the total number of subunits in the base. The sum of the total number of subunits is always equal to unity since the base of the model square is always equal to unity.

For example:

A. |------------------------------------|

Undivided base of model square equal to 1

In example A, the base of a model square is always equal to 1.

```
      1   2   3   4   5   6   7   8
B. |----|----|----|----|----|----|----|----|
```

Divided base of model square = 8

In example B, the base of the model square in example A is divided into 8 subunits of 0.125=1; (1/b)

In the fraction 1/b in example B, the divided base of the model square is always equal to the denominator b. The denominator of the fraction, b, in this case is equal to 8, and is also the factor by which the base of the model square is divided; but at the same time, it also represents the number of subunits in the model square.

```
        1       2       3       4
C. |---- ----|---- ----|---- ----|---- ----|
```

Divided base of another model square
Divided base of model square = 4

In example C, the base of the model square in example A is divided into 4 subunits of 0.25 = 1; (1/b)

Values are not drawn to scale

Examples shown in A, B, and C illustrate how measurements can be different for the value of 1 when this value is divided into subunits. It is this relationship between values that is applied to the ratio concept in relation to geometric figures.

In figure A, the base of the model square is equal to unity (base = 1) is undivided. This square is not a monad. In the fraction $1/b$, $b = 1$; the base of the model square is represented by the fraction $1/1 = 1$. Since the quotient of $1/1 = 1$ is a whole number, the model square is not seen as a monad. A monad is always fractional and this is why b must always be greater than 1 to express the side of a monad.

Once the base equal to unity is divided, it becomes the divided base. The divided base is made of an equal number of subunits that is equal to unity, and $1/b$ becomes the ultimate unit of measurement for the model square; in addition, $(1/b)^2$ is equal to the area of the ultimate monad for this particular model square. The same reasoning can be applied to the model rectangle and the model circle.

When we talk about the base of the Model square, model rectangle, or model circle, I am usually referring to the *divided base*. (The author wants to underline one more time that the base concept may not be important in the case of a square, but it is very important in the case of a rectangle and a circle. I would like to underline one more time that the radius is considered as the base, when we refer to the circle's base.)

Model squares, model rectangles, and model circles have certain properties. If we calculate the dimensions of a model square according to its ratio, the ratio of the model square is always equal to its area. When we calculate the dimensions of a model rectangle according to its ratio, its ratio is always equal to its area. The same is valid for the model circle; its ratio is also equal to its area.

Model rectangles and model circles computed from the same fraction and ratio, have the same area; furthermore, the added lengths of a model rectangle is equal to the circumference of a model circle of the same ratio, and the added widths of the model rectangle is equal to the diameter of the model circle. These are things that we never understood before and never accomplished before.

When we divide the base of a model square by $1/b$, we are automatically modifying the internal structure of the model square of base 1 since the side of the monad is equal to $1/b$, and the area of each individual monad is equal to $(1/b)^2$. As we divide the base of the model square by larger and larger numbers using the formula $1/b$, the magnitude of the internal squares get smaller and smaller, and they can reach a point where a column made of squares becomes so thin that it could almost coincide with the lines forming the sides of a square. As our calculations approach infinity, we cannot help thinking that a column made of such minute squares could reach atomic proportions. The thickness of the lines made of these squares could be thinner than a line made by a sharp pencil.

Theoretically, a column or straight line made of squares that are so minute could be used to create the area of an approximate circle by drawing first the diameter, and then adding shorter adjacent and parallel lines to the diameter with the extremity of each line stopping at the perimeter of the circumference of the circle.

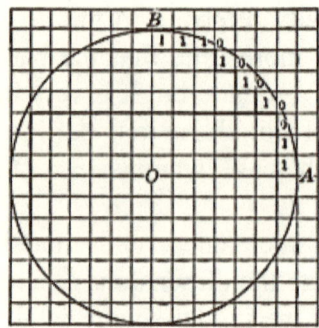

Please note that the product for the calculation of squares can move toward the infinitely large or the infinitely small.

For example, if we consider the following squares $2^2 = 4$, $3^2 = 9$, $4^2 = 16$ . . . and so on, the area of these squares is moving toward the infinitely large. When the product moves toward the infinitely large, $2^2 = 4$ means we are dealing with 4 square units equal to 1^2 each in area, and each of these units represents a model square unit whose side is equal to unity. Please note that this square unit equal to unity does not have a divided base and is not fractionalized. It is used in calculations where the product of numbers moves toward the infinitely large.

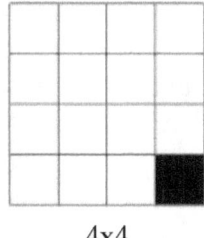

4x4

The dark unit in the 4x4 square above is a model square. This model square is what the magnitudes having areas moving toward the infinitely large are made of, and they are all equal to each other. *These square units are model squares with an undivided base (1 x 1).*

When the product moves toward the infinitely small, $(\frac{1}{2})^2 = \frac{1}{4} = .25$; $(1/3)^2 = 1/9 = 0.11111 \ldots$; $(\frac{1}{4})^2 = 1/16 = 0.0625$; $(1/5)^2 = \ldots$, we are strictly fractionalizing the internal structure of one model square whose base is equal to unity.

(The small dark square of 1x1 from the picture above is magnified below and could be divided into smaller squares of 4x4 or 5x5).

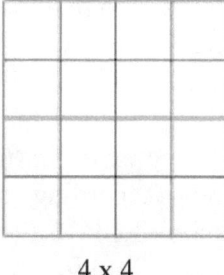   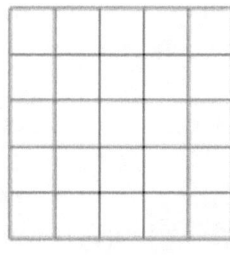

4 x 4   5x5

These squares are the monads and their size is smaller than unity. The side of the monad is equal to 1/4 for square 4x4. The side of the monad is equal to 1/5 for square 5x5. The total area of all the monads in each square is always equal to 1.

These model squares have a divided base.

The smaller squares resulting from the divided base of the model square whose base is equal to unity are the monads. A monad can always be differentiated from a model square with an undivided base. The area of the model square with an undivided base is always equal to 1, and the area of a monad is always fractional. In addition, the magnitude of the monads are always smaller than unity, and their area always move toward the infinitely small and not toward the infinitely large as b gets larger in the fraction $1/b$. The length of the side of a single monad is the ultimate unit of measurement to calculate the perimeter of a model square, and the area of a single monad in the model square represents the ultimate square unit that makes the area of the model square. The dimensions of a single monad represent the ultimate standard of measurements that will satisfy a particular model square with a very specific ratio.

The example of the model square was used as an example, but the same type of reasoning will also apply to the model rectangle and the model circle.

(Note: *Reverse multiplication*: A term proposed by the author to indicate that the product resulting from a multiplication operation is smaller than the multiplicand, as in the case of proper fractions (e.g., ¼ x ¼ = 1/16). As a reminder, a proper fraction is a fraction that the numerator is smaller than the denominator.

Technically for centuries, the scholars have applied the terminology "multiplication" erroneously to define the *reverse multiplication* that takes place in the case of proper fractions. According to the rules of multiplication, if we multiply a number such as 2 x 3 = 6, the multiplicand (2) can be added three times by the multiplier (3) to find the product (6), that is—2 + 2 +2 added three times is equal to 6.

But in the multiplication of proper fractions such as ¼ x ¼ = 1/16, the product of the numbers being multiplied move toward the infinitely small, as opposed to a multiplication of whole numbers where the product move toward the infinitely large; and the multiplicand cannot be added a certain number of times to find the product since the product is smaller than the multiplicand; this is proof that a multiplication was not performed in the case of proper fractions. It is obvious that the modus operandi for reverse multiplication does not follow the modus operandi of multiplication for whole numbers, and new criteria must be established in the case of reverse multiplication.

Therefore, in all humility, not only a new terminology is necessary in the case of multiplication of proper fractions; but also a new sign other than "x" or "·" may be needed to indicate a new operation. There is a great difference between operations resulting in a product that moves toward the infinitely large and one that moves toward the infinitely small. It is obvious that proper fractions do not follow the same rules of multiplication as whole numbers. We should provide and suggest new criteria for the reverse multiplication that takes place in the calculation of proper fractions.)

The perception of reverse multiplication is important because when we divide $1/b$, the dimensions of the monads also reverse direction, and the measurement of the side of the monad and the area of the monad get smaller as b gets larger; in other words, we have multiplied, but the dimensions get smaller and do not follow the criteria conforming to the rules of multiplications.

In the next paragraph, we are going to determine the shape of a square according to its ratio. In addition, we will try to find the dimensions of its individual monads and the total number of monads representing the square; furthermore, we will determine its area and verify it against its ratio.

Objective 9. Calculating the area of a model square and a model rectangle according to their ratio based on the concepts of the monad conjecture

The reader will realize that it is the same *Fabius Universal Ratio Formula* invented by the author $a/b = (1/b)^2 (a \cdot b)$ that was previously used to calculate the numbers of squares in a ratio in the fraction $a/b = q$ that is now modified to calculate the area of each geometric figure in accordance with their particular ratio.

If we replace the letters in our universal ratio formula by $h/b = 1$; the sides of a square being represented by h for height and b for base and the square being considered as a rectangle of equal sides, so $h/b = 1$ becomes $h/b = (1/b)^2 (h \cdot b)$;

($h = b$ in the case of a square and remember b is always equal to the divided base)

Note: the concept of height and base being equal is more important than the s^2, side multiplied by side concept used in the calculation of the area of squares, because the height being perpendicular to the base automatically implies an existing right angle that the concept of s^2 does not include; and this idea is very important to express the concept of a square in terms of numbers.

Properties of a model square:

For the purpose of explanation and clarity only, the *modified Fabius Universal Ratio Formula* for the calculation of dimensions of model squares according to their ratio at a glance is:

$h/b = (1/b)^2 (h \cdot b)$; (h=b in the case of a square)

This formula determines the area and the ratio of a model square, which is always equal to 1.

Proposition 1: The base of a model square is always equal to unity. The divided base of a model square is always equal to its dividing factor b, and b always represents the number of subunits pertaining to the base of a model square. The side of a single monad in a ratio is equal to 1/b and also represents the ultimate element of measurement or physical reality for this particular model square, and this measurement cannot be further divided by definition. The number of subunits in the divided base multiplied by the length of each individual monad is equal to the length of the base which is equal to unity, $b \cdot 1/b = 1$.

Proposition 2: The ratio of the sides (h/b) of a square is always equal to unity.

Ratio: h/b $= 1$ - *(no demonstration)*

Proposition 3: The ratio and the area of a model square is always equal to 1.

h/b = (1/b)² (h · b) = *area = ratio*

Proposition 4: The perimeter of a model square is equal the total number of subunits that compose its base (b) times the number of sides of the square (4) multiplied by the side of each individual monad (1/b).

The perimeter of a model square can be computed with this formula:

(b· 4) (1/b) *and is always equal to 4. (no demonstration)*

Calculating the dimensions of a model square using the universal ratio formula:

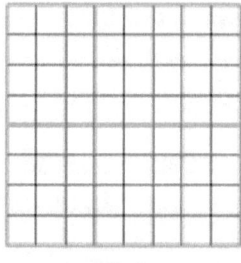

8x8

Divided base equal 8

In the square above, if *h* = 8 and *b* = 8, find the internal dimensions of a model square of ratio 8/8 = 1 according to the *Modified Fabius Universal ratio formula*:

h/b = (1/b)² (h · b)

The ratio of the square is *h/b = 1* or 8/8 = 1; the ratio 1 determines the shape of the square since the ratio of the sides of all squares is equal to 1. The ratio 1 is also equal to the area of the square.

The *modified Fabius universal ratio formula* at a glance is:

h/b = (1/b)² (b · h) ; or,

$8/8 = (0.125)^2 (8 \cdot 8)$; or,

$8/8 = 0.015625 \times 64 = 1$

The explanation of the above equation is found below:

$1/b$ = formula to calculate the side of the monad:
$1/8 = 0.125$

$(1/b)^2$ = formula to calculate the area of the monad:
$(1/8)^2 = 0.015625$

$h \cdot b$ = formula to calculate the number of monads in a ratio
$(8 \cdot 8) = 64$

$(1 \cdot b)^2 (h \cdot b)$ = Formula to calculate the area or the ratio of the parted model square:
$0.015625 \cdot 64 = 1,$

The number of squares and their individual dimensions are verified against the area of the model square and its ratio for fraction $8/8 = 1$.

The monads can be grouped together, at least in one instance, to form a model square. The results show that 64 monads having an individual area = 0.015625 are equal to an area of 1. The ratio of its sides, height/base, is equal to 1.

The side of the model square is also equal to the number of subunits in the divided base multiplied by the side of the monad:

Base of the model square = $b (1/b) = 8 \cdot 0.125 = 1$

The formula for the perimeter of the model square is:

$(b \cdot 4) (1/b) = 4$ (the perimeter of a model square is always equal to 4)

Calculating the dimensions of a model rectangle

2) In the same order of ideas, in the fraction a/b= q, the monads can be grouped together to represent a rectangle if the ratio q of their sides is different from 1. In this case, we are fractionalizing the base of a rectangle equal to unity into smaller squares (monads). A square is viewed as a rectangle of equal sides, and a rectangle simply as having unequal sides. We can demonstrate that, at least in one instance, the total value of the squares contained in a ratio can be arranged as a rectangle, if the ratio is different from 1, ($h/b \neq 1$). For calculation purposes, just like the square, the numerator of the fraction, a, represents the height of the rectangle; and the base of the rectangle represents the denominator, b. Side a is always different from b. Dividing side a by side b finds the ratio. Using the base and the height for the rectangle to replace the formula ($l \cdot w$) or (length multiplied by width) implies a ninety-degree angle.

The ratio of $h/b \neq 1$ (h/b different from 1) is symbolic of a rectangle.

Similarly for a rectangle, we can replace the fraction $h/b \neq 1$ by the *modified Fabius universal ratio formula* invented by the author to calculate the dimensions of a rectangle:

$h/b = (1/b)^2 (h \cdot b)$; where $h \neq b$; *(if h is different from b, the ratio will always be different from 1)*

Basic explanations for the calculation of rectangles according to their ratio using the monad conjecture approach

Please note that using the denominator as the base does not make much difference in the case of a square, but it makes a huge difference in the case of a rectangle. The idea of position becomes more important in the case of the rectangle since either the length or the width of a rectangle can be used as a base. In the monad conjecture, the dimensions of the individual squares (monads) are computed strictly according to the *base* of the model rectangle. Therefore, the ratio will be different, and the dimensions of the monad will vary depending on which side is used as a base. If the dimensions of the monad vary, the dimension of the whole rectangle will vary. In Euclidean geometry, it makes no difference which side is used to compute the area because the unit of measurement is based on 1; but in the monad conjecture, the base is always a fraction of 1 and we use the subunit of the divided base as the ultimate unit, so we must always be aware of the dimensions of the unit we are using since it may affect the whole dimension of the rectangle.

h=height   b=divided base

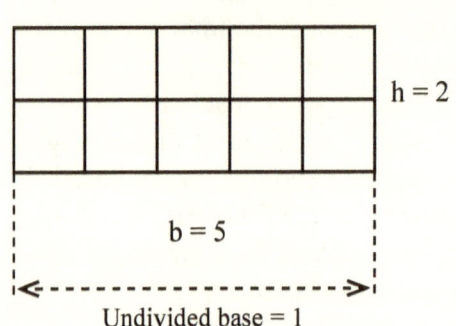

h = 2

b = 5

Undivided base = 1

Ratio = 2/5
Undivided base = 1
Divided base = b = 5
Side of monad = 1/5

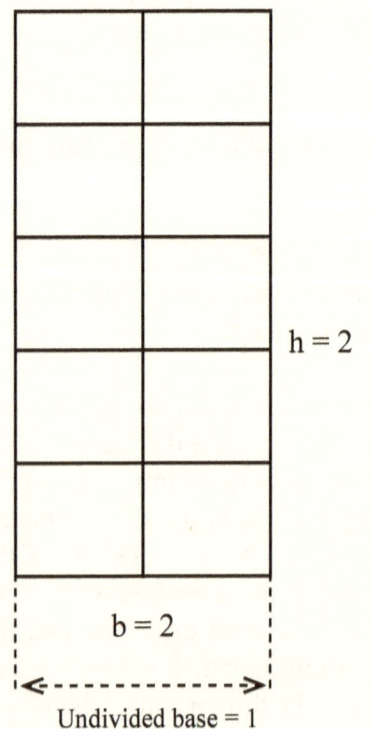

h = 2

b = 2

Undivided base = 1

Ratio = 5/2
Undivided base = 1
Divided base = 2
Side of monad = 1/2

Not drawn to scale

In this context, a rectangle with a ratio of 2/5 has the same number of squares than a rectangle of 5/2; but the ratios are different, the individual squares have different dimensions, and the rectangles have different dimensions. Remember the total sum of the area of the individual squares must be verified against the ratio of the rectangle. This perception is very important because the dimensions vary with the ratio, and when we are working with two different bases, we are not working with the same unit of measurement. A conversion of the measurements would be necessary in order for the area and the perimeter of the rectangles to coincide! Only one side of the rectangle can be used as a base.

Properties of a model rectangle:

For the purpose of explanation and clarity only, the *modified Fabius Universal Ratio Formula* for the calculation of dimensions of model rectangles according to their ratio at a glance is:

$h/b = (1/b)^2$ ($h \cdot b$) also = **ratio** = **area;** *but in this case, the ratio is always different from 1 since the sides are always unequal.*

Proposition 1: The base of a model rectangle is always equal to unity. The divided base of a model rectangle is always equal to its dividing factor, b, *and* b *always represents the number of subunits pertaining to the base of a model rectangle. The side of a single monad in a ratio is equal to 1/b and also represents the ultimate element of measurement for a particular model rectangle, and this measurement cannot be further divided by definition. The number of subunits in the divided base multiplied by the length of each individual monad is always equal to the base of the model rectangle that is equal to unity, b·1/b=1.*

Proposition 2: The ratio of the sides (h/b) *of a model rectangle is always different from 1.*

$h \neq b$

$h/b \neq 1$ *(ratio always different from 1)*

Proposition 3: The area of a model rectangle is always equal to its ratio.

$h/b = (1/b)^2 \, (h \cdot b)$

Proposition 4: Model rectangles and model quasi-circles having in common the same base, the same fraction, and the same ratio are also equal in area. (This proposition will be demonstrated after the reader gets acquainted with the quasi-circle theory.)

Proposition 5: If a model rectangle and a model quasi-circle have in common the same base, the same fraction and ratio, the perimeter of the rectangle is equal to the perimeter of the circle plus its diameter. (This proposition will be demonstrated after the reader gets acquainted with the quasi-circle theory.)

Proposition 6: If a model rectangle and a model quasi-circle have in common the same divided base and the same fraction and ratio, the perimeter of the circle is equal to the perimeter of the rectangle minus both widths. (This proposition will be demonstrated after the reader gets acquainted with the quasi-circle theory.)

Proposition 7: If a model rectangle and a model quasi-circle have in common the same fraction and ratio, the perimeter of the circle is equal to both sides that make the length of the rectangle, and its diameter is equal to both widths of the rectangle.

Proposition 8: The perimeter of a model rectangle is equal to twice the length of its base multiplied by the length of the side of one single monad plus twice the length of its widths multiplied by the length of one single monad.

The formula for the perimeter of a model rectangle at a glance is:

$[(2b) \, (1/b)] + [(2h) \, (1/b]$

Determining the number of monads in a rectangle according to the ratio of its sides and verifying the number of squares against the ratio by using the universal ratio formula:

Let's calculate the ratio and area of a rectangle of height, 2, and base, 5, according to the Fabius Universal Ratio Formula.

a/b =(1/b)² (a · b)

In the original universal ratio formula, let's replace *a* by the height (*h*) of the model rectangle and *b* remains *b* for the base to represent the sides of the rectangle.

For the purpose of explanation and clarity only, the *modified Fabius universal ratio formula* for the calculation of dimensions of model rectangles according to their ratio at a glance is:

h/b=(1/b)² (h·b)

The ratio of the sides of the rectangle *h/b* (height/base) is:

$2/5 = 0.40$

Solving the equation:

$2/5 = (1/b)^2 (h \cdot b)$

$2/5 = (1/5)^2 (2 \cdot 5)$

$2/5 = (0.20)^2 (10)$

$2/5 = 0.04 \cdot 10$

$2/5 = 0.40$

Explanation of each individual formula that makes up the equation:
Formula to calculate the side of the monad:

$1/b = 1/5 = 0.20$

Formula to calculate the area of each individual monad:

$(1/b)^2 = (1/5)^2 = 0.04$

Formula to calculate the number of monads in the rectangle:

$h \cdot b = 2 \cdot 5 = 10$ or number of monads in the rectangle

Formula to calculate the area of the rectangle

$(1/b)^2 \ (h \cdot b)$
$0.04 \cdot 10 = 0.40 =$ area of model rectangle $=$ ratio

The 10 monads with an individual area of 0.04 can be grouped together to form the area of a rectangle equal to 0.40, at least in one instance. The ratio is also equal to the area of the rectangle.

Sides of the rectangle can be found by multiplying the number of subunits in the divided base by the side of each individual monad and the height multiplied by the side of each individual monad.

Base of rectangle $= b \ (1/b) = 5 \cdot 0.20 = 1$
Height $= h \ (1/b) = 2 \cdot 0.20 = 0.40$

The perimeter of the rectangle is:

$[(2b) \ (1/b)] + [(2h) \ (1/b)]$

Please realize that the concept is the same when we are dealing with a rectangle as when we are dealing with a square, we are fractionalizing the internal structure of the rectangle. The number of squares and their dimensions are verified against the ratio of the rectangle, and the ratio represents the area of the model rectangle.

The same concept also applies to quasi-circles. It is the same formula that is modified to calculate the area of circles, but in the case of the circle, the inner dimensions of the circle have never been calculated before.

Objective 10. Exploring basic mathematical concepts behind the quasi-circle theory and calculating quasi-circles (subdivided into eighteen objectives)

The quasi-circle theory is an integral part of the monad conjecture, and the author suggests that the reader get acquainted with it first before proceeding with our study of the monad conjecture. The study of approximate circles is inextricably tied to the monad conjecture but can be considered altogether as two separate studies.

(Please note that objectives for the monad conjecture are in numerical sequence, and objectives for the quasi-circle theory are in alphabetical sequence.)

What is the quasi-circle theory?

The quasi-circle theory is the study of approximate circles. It is also the study of the circle seen from a numerical point of view.

In this chapter, the reader is invited for a brief overview of a parallel concept to calculate the area of circles that the author developed during the study of quasi-circles. In this study, the dimensions of approximate circles are calculated according to their ratios. The same universal ratio formula used to calculate the square and the rectangle in our study of the monad conjecture was modified and adapted for the calculations of the circle. These calculations of the dimensions of the circle were never accomplished before this work.

Before to continue, I would like to establish some of the facts for a model quasi-circle of unit radius:

In the quasi-circle theory, the formula to calculate the number of squares in a ratio is still the same universal ratio formula that has been modified already twice for the calculations of the model square and the model rectangle and is now being modified one more time to calculate the area of model circles. The *Fabius universal ratio formula*, one more time, is:

$$a/b = (1/b)^2 \, (a \cdot b)$$

At a glance, here is the modified Fabius universal ratio formula to calculate the area of a model quasi-circle. In this formula, we have changed a/b by $\frac{1}{2}\,C/R$; $a = \frac{1}{2}\,C$ (C = approximate circumference) and b has been changed to R (divided base of the radius):

$$\tfrac{1}{2}C/R = (1/R)^2\,(1/2C \cdot R)$$

Proposition 1: The base of a model circle is always equal to unity, and the radius represents the base of a model quasi-circle. The divided base of a model circle is always equal to its dividing factor R, and R always represents the number of subunits pertaining to the base of a model quasi-circle. The side of a single monad in the ratio of a quasi-circle is equal to 1/R and also represents the ultimate element of measurement for a particular model circle, and this measurement cannot be further divided by definition. The number of subunits multiplied by the length of a single monad is equal to the base(radius) of a quasi-circle which is equal to unity, R·1/R.

Proposition 2: The ratio of a quasi-circle always begins with the first three digits of pi, 3.14 . . . , and is always a rational number represented by a fraction. (See objective D, rationalization behind using any ratio starting with 3.14 . . . as the ratio of all circles.)

Proposition 3: The ratio of a circle of unit radius is always equal to its area or ½ its perimeter.

(Propositions 4 to 6 will be demonstrated as we pursue our study of the monad conjecture after the reader gets fully acquainted with the quasi-circle theory.)

Proposition 4: Model rectangles and model quasi-circles that have in common the same divided base, and the same fraction and ratio are also equal in area.

Proposition 5: If a model rectangle and a model quasi-circle have in common the same divided base and the same fraction and ratio, the perimeter of the rectangle is equal to the perimeter of the circle plus its diameter.

Proposition 6: If a model rectangle and a model quasi-circle have in common the same divided base and the same fraction and ratio, the perimeter of the circle is equal to the perimeter of the rectangle minus both of its widths.

Proposition 7: If a model rectangle and a model quasi-circle have in common the same fraction and ratio, the perimeter of the circle is equal to both sides that make the length of the rectangle, and its diameter is equal to both widths of the rectangle.

I promise to the reader that the definition of the word "area" as currently defined in the dictionary will not be altered, and it will be met as the quantitative measure of a plane or curved surface as perceived in a two-dimensional extent.

After all the objectives for the quasi-circle theory are met, we will proceed to show how a rectangle and a quasi-circle of the same ratio are equal in area; and since all rectangles can be squared, therefore, all quasi-circles can be squared.

Objectives of the quasi-circle theory:

A. *Provide a geometric definition for quasi-circles*
B. *Provide a numerical definition for quasi-circles*
C. *Outline the goals of the quasi-circle theory*
D. *Discuss the rationalization behind using 3.14 . . . as the ratio of all circles, including Quasi-circles*
E. *Establish the criteria to Identify quasi-circles*
F. *Compute a quasi-circle fraction for a circle*
G. *Calculate the accuracy of quasi-circles*
H. *Define concentric and eccentric quasi-circles*
I. *Point out that old fractions have now new meanings and functions*
J. *Provide a formula to determine the side of each individual monad (individual square) that makes the area of a quasi-circle*
K. *Provide a formula to determine the area of each individual monad that makes a quasi-circle*

L. *Provide a formula to find the precise number of individual squares that makes the area of approximate circles using any fractions within the quasi-circle range of ratios*

M. *Provide a Formula to find the precise area of a quasi-circle according to its ratio*

N. *Validate the accuracy of the area of a model quasi-circle against its ratio*

O. *Calculate the number of rectified arc units that makes the perimeter of a circle*

P. *Find the precise length of each rectified individual arc unit based on the ratio*

Q. *Calculate the precise length of the perimeter of the circumference of a quasi-circle according to its ratio*

R. *Validate the degree of accuracy of the circumference of the circle against its ratio*

S. *Provide a formula to compute the expression of roundness of a circle based on the old polygon concept of Archimedes*

Objective A. Provide a geometrical definition for quasi-circles

As defined in the first chapter, a quasi-circle is a circle that is slightly disproportionate from its diameter. Another name for a quasi-circle is an approximate circle. Theoretically, the area of a quasi-circle is made of a specific number of individual squares, tightly packed together, ranging from small to subatomic dimensions that are arranged in a circular pattern to create the illusion of a perfect circle.

The quasi-circle theory uses a polygon concept to show how polygons can create the illusion of a perfect circle:

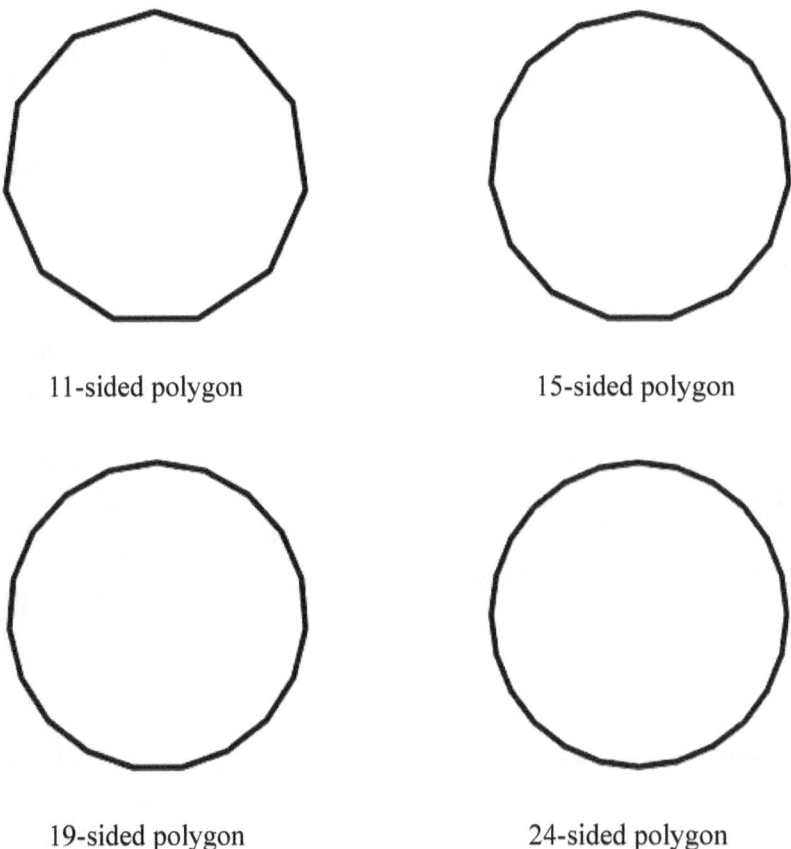

11-sided polygon   15-sided polygon

19-sided polygon   24-sided polygon

Quasi-circles exist within a specific range of ratios, and their numbers are infinite. The dimensions of quasi-circles can get so close to those of the perfect circle that their differences may not be perceptible through the naked eye, but may be highly perceptible through calculations. For the purpose of geometry, we may substitute a perfect circle for a quasi-circle.

The advantage of quasi-circles is that it allows for very precise calculations in regards to the internal structure of the circle, and we would never be able to achieve this kind of precision by using strictly a transcendental number.

In the monad conjecture, we have shown how the internal structure of a square equal to 1^2 in area could be divided indefinitely until the columns containing the squares become so thin that they could almost coincide with the lines that bound the square. As I mentioned in a previous chapter, physicists have determined that it will take approximately twenty million hydrogen atoms to make a unit as long as this "-". Imagine that we can calculate squares that are much smaller than the thickness of the straight line that you perceive in this "-".

Quasi-circles are represented by fractions. As I have stated before in previous paragraphs, the numerator and denominator of a fraction can be multiplied indefinitely by the same number, and the fraction itself could reach astronomical numbers while conserving the same ratio. The larger the denominator in the fraction, the smaller will be the individual squares. There is overwhelming mathematical evidence in this work that a quasi-circle can mimic the perfection of a perfect circle.

$n(a)/n(b)$ = QCR (quasi-circle ratio)

Objective B. Provide a numerical definition for quasi-circles

Theoretically, a quasi-circle, from a numeric point of view, is an approximate circle that could be created from specific dimensions computed from a quasi-circle fraction. A quasi-circle is always represented by a quasi-circle fraction which always starts with 3.14 . . . , and this number automatically establishes the range of ratio for quasi-circles. (See objective D to find the study behind the rationalization behind choosing 3.14 . . . as the ratio of all circles.)

Before creating a quasi-circle, it is best to determine first a quasi-circle fraction to represent the quasi-circle (see objective F on calculating a quasi-circle fraction). The ratio from the quasi-circle fraction can be compared to pi to determine the desired accuracy (see objective H on calculating the accuracy of concentric quasi-circles and eccentric quasi-circles). The quasi-circle fraction contains all the elements necessary to calculate the specifications of an approximate circle based on a circle of unit radius.

A circle with a parted radius (divided base) equal to unity is a model circle. Once the dimensions of the model circle have been determined, then other circles of greater diameter can be calculated using the same quasi-circle ratio. Once we have chosen a ratio, this ratio represents our equivalent of pi, and the same ratio must be used throughout all our calculations. We have dealt with different ratios before to represent pi, but we never had a full comprehension of their applications.

Objective C. Outline the goal of the quasi-circle theory

The purpose of the quasi-circle theory is to offer an alternate method to calculate the dimensions of circles by providing a formula that would enable engineers, scientists, and theorists, to calculate a measurable version of pi (MVP), which would allow them to perform better in their respective professional capacities. This measurable version of pi (MVP) contains the specifics for a particular ratio that can be seen as the genetic equivalent of the ratio (see the specifications below for this particular quasi-circle ratio); this amount of information was not available to them before this work, at least not with precision. Professionals in their respective line of work can create their own measurable version of pi depending on the degree of accuracy they wish to achieve. The user of the ratio would also be aware of the level of accuracy offered by the ratio he has decided to use.

This measurable version of pi can be integrated in a scientific calculator, a programmable calculator, or a computer. (This is why the author is announcing his intent to patent the *measurable version of pi* for electronic devices, especially those with radius or diameter = 2^n that meet both requirements for quasi-circles. The author also reserves the right to manufacture his own calculator with MVP and also the right for leasing MVP for use in electronic devices.)

An engineer may find it more appropriate to have his own version of pi if he wants to build a round computer chip that contains a large amount of minute squares. He may even use my own measurable version of pi that is equal to 3217/1024 = 3.1416015625. This version of pi is called a quasi-circle ratio and contains the following specifications:

(see next page)

(Please note that these are exact dimensions)

Specifications for a quasi-circle of unit radius computed from a quasi-circle fraction equal to 3217/1024:

½ C/R = 3217/1024

C = 2 (½C) = 6434 (number of rectified arc units that makes the circumference of a circle of unit radius)

R = 1024 (number of units equal to the divided base of the radius)

Radius: r = 1

Diameter: d = 2

Quasi-circle ratio: QCR = 3.1416015625

Level of accuracy = +0.000008909 (eccentric quasi-circle)

Length of the side of each individual monad = 0.0009765625. (Length of the side of each monad is also the length of each arc unit measured in a straight line that makes the circumference for this particular ratio.)

Perimeter of circle of unit radius:
6434 · 0.0009765625 = 6.283203135

½ Perimeter of circle of unit radius:
3217 · 0.0009765625 = 3.1416015625

Area of each monad = 0.00000095367431640625

Number of monads (individual squares) that make the area of this quasi-circle = 3,294,208 monads

Area of circle:
3,294,208 · 0.00000095367431640625 = 3.1416015625

Sides of model rectangle equal to this model circle of unit radius:
(length) 3217 (0.0009765625) · (width)1024 (0.0009765625)

Expression of roundness:
$(1/(1024)^2 = 0.00000095367431640625$

The engineer now has all the information available about this version of pi that he is about to use to build a computer chip—information he would have never been able to get if he was using our present method of calculation. Since the diameter is made of geometric dimensions, he can also achieve better accuracy during the manufacturing process of this chip.

Objective D. Discuss the rationalization behind using 3.14 . . . as the ratio of all circles including quasi-circles

There are two major reasons why any ratio starting with 3.14 . . . was established as the ratio of all circles in the quasi-circle theory. One is historical, and the other one is academic.

When Archimedes, the famous Greek mathematician who lived around 255 BC, calculated the ratio of the circumference of the circle to its diameter to be less than 3 1/7 (22/7) and greater than 3 10/71 (223/71), he established the limits of pi to be, in decimal equivalents, between 3.142857 . . . and 3.140845 . . . Since that time we have never stopped calculating pi to obtain new limits of accuracy for the number.

From an observer's point of view, if we take a retrospective look into mathematical history, it will tell us that each of the following famous mathematicians calculated their own version of pi; and they all consistently found 3.14 . . . in their computation for the first three digits of pi. Ptolemy (ca. 150); Liu Hui (ca. 263); Zu Chongzi (ca. 470); Āryabatha (ca. 510); Al-Kwarizmi (ca. 800); Franco of Liege (ca. 1066); Fibonacci (ca. 1220); Madhava (ca. 1400); Jamshid Al-Kashi (ca. 1430); Viete or Vieta (ca. 1593); Ludolf Van Ceulen (ca. 1596); Aedriaen Van Roomen (ca. 1561-1615); Newton (ca. 1665); Georg

Vega (ca. 1794), who calculated pi to 140 decimal places and had 126 correct; Zaharia Dase (1824-1861), who carried pi to 200 decimal places; William Rutherford, who found pi to 200 places by using John Machin's formula; William Shanks, who calculated pi to 707 places of which 527 were correct; and countless other mathematicians not named in this paragraph, who deserve credit as well, also confirmed the parameters of at least the first three digits of pi to be 3.14 . . . as established by Archimedes since 225 BC.

Since Franco of Liege (ca. 1066), no questions have been raised regarding the accuracy of the first three digits of pi being different from 3.14 . . . by any reputable mathematician in the world. These numbers can be verified in my chapter, "History of Pi," for all the aforementioned mathematicians.

In an academic sense, it appears that a clear consensus has been reached unconventionally by mathematicians all over the world indicating that pi represents the approximate ratio of the circumference of a circle to its diameter; and these concepts reflect a rigid order of numbers in their computation of pi throughout history.

When the English writer William Jones first expressed pi as the ratio of c/d in his *Synopsis Palmariorum Matheseos* in 1706, it was not known whether pi was rational, irrational, or transcendental. The quasi-circle theory had not entered the picture of the circle as yet, so no one questioned the validity of the statement c/d = pi; c/d is rational, and pi cannot be represented by c/d since it is an irrational number. If there is a way to make an expression equal to something else, mathematicians will find it, and if they can't find it, they will create a new way; so they made pi an approximation of the ratio of c/d.

Mathematicians who expressed pi as the quotient of c/d after the proof of the transcendence of pi given by the famous German mathematician Ferdinand von Lindemann in 1882 simply lack mathematical integrity since they are very well aware that pi cannot be represented by a fraction. Pi stands alone and cannot be represented by a fraction. This is why the author invented a new terminology, quasi-circle ratio, which is rational and always represented by a fraction, to express the relation between the circumference of the circle and its diameter.

Historical evidence shows that scholars from all over the world have computed and agreed on the sequential order of pi. For instance, when William Shanks calculated pi to 707 digits, mathematicians proved that only 527 were correct. While pi is considered an approximate ratio, it is also the most extensively calculated number by mathematicians in the history of the world. Fabrice Bellard of France has now calculated pi to over 2.7 trillion decimal digits, and new digits are constantly being calculated and added to the number. The reader can refer to my chapter in this book titled "History of Pi" to get a better idea of how pi evolved from an approximate ratio of 3 during Babylonian and Egyptian times to become a transcendental number computed beyond 2.7 trillion digits in modern times.

Thus, the range of QCR (quasi-circle ratio) starting with 3.14 . . . is not arguable numerically. Since a quasi-circle ratio is a derivative of pi and can be expressed by any fraction that represents pi taken to any decimal digit extent, it is safe to say that the ratio of all circles, perfect or approximate, starts with 3.14 . . . based on the work of well-known mathematicians throughout history. It could have been very easy to use the parameters established by Archimedes, but 3.14 . . . seems to be more appropriate to the space age and automatically establishes a wide range for quasi-circle ratios at a glance without having to perform any new calculations. Besides, 3.14 . . . includes the ratio 22/7 found in the works of Archimedes even though this ratio probably goes further back than Archimedes in history, but at least, using such a range of ratios gives credit to the early mathematicians for their work on the circle.

Since pi has already been calculated beyond 2.7 trillion decimal digits, there is absolutely no need to calculate pi one more time to arrive at our calculations. The author provided pi beyond thirty digits after the decimal points in one of the following paragraphs for the purpose of calculating quasi-circle fractions, and should the reader need more accuracy, he can use any book reflecting the computation of pi to a zillion decimal digits and use it to calculate quasi-circle fractions. Once the quasi-circle fraction is determined, this fraction contains all the elements necessary to calculate the dimensions of a quasi-circle.

Quasi-circle fractions reduced to their simplest expression are never expressed as the ratio of c/d but as the ratio of $\frac{1}{2}C/R$ ($\frac{1}{2}$ the *circumference/*

radius). The expression of the formula has changed, but the ratio stays the same. However, when we double a quasi-circle fraction from its simplest expression, it can be represented as the symbol of *c/d* (*circumference/ diameter*), but in the quasi-circle theory, it has a different meaning. The reason for changing the formula from *c/d* to ½ *C/R* is to maintain uniformity with the previous concepts taught in the monad conjecture and to avoid the fragmentation of the monad, which cannot be further divided by definition according to the monad conjecture when working with a particular ratio. Mathematicians obviously do not see themselves as architects or engineers and completely disregarded the ultimate building block concept that represents the foundation of all things. Of course, there is always a price to pay for a fallacy in logic.

For the purpose of this work, pi is considered as the best approximate ratio of the circumference of the perfect circle to its diameter, and this ratio is used as a guide to find quasi-circle fractions. This is why the author said in the first chapter that pi fits better the quasi-circle theory than the perfect circle concept, and its function has become much more extended than the unique role of approximate ratio that it plays in the calculations of perfect circles because it has now become a guide in the calculation of quasi-circle fractions. Nevertheless, pi is not a quasi-circle ratio and does not meet the requirements for quasi-circles ratios. Once more, pi is only considered as a guide to find quasi-circle fractions.

Pi = 3.14159265358979323846264433832795 . . .

Examples of quasi-circle ratios:
½ C/R = QCR

22/7 = QCR = 3.142857 (Archimedes)
355/113 = QCR = 3.1415920 (Zu Chongzi)
3217/1024 =QCR= 3.1416015625 (Lionel Fabius)
105414357/33554432= QCR = 3.14159265160 (Lionel Fabius)

The digits in bold letters represent the level of accuracy for these quasi-circles ratios in comparison to pi.

Pi is the approximate ratio of the circle to its diameter. It is a transcendental number and cannot be expressed as the ratio of two integers. A quasi-circle ratio starts with 3.14 . . . and is always different from pi in the sense that pi has no limits, but a new quasi-circle fraction and ratio can be calculated ad infinitum for every new limit of pi, and this fraction contains all the elements to calculate the dimensions of a quasi-circle. The ratio of an approximate circle is always a rational number and can always be represented as the ratio of two integers.

Objective E. Identifying quasi-circles

From a mathematical standpoint, an approximate circle must meet two of the following criteria in order to qualify as a quasi-circle:

1. The ratio of ½ its circumference to its radius ($\frac{1}{2}C/R$), must be always a rational number represented by a fraction made of positive integers. *(C, in this case, represents the number of rectified arc units that makes the circumference of a Quasi-circle of unit radius and R the number of subunits that makes the radius of a quasi-circle.)*

For example:
½C/R = QCR (quasi-circle ratio)
22/7= 3.1428571

2. The ratio or quotient of the fraction must always be a derivative of pi and must always begin with 3.14 . . . . This number automatically establishes the range of ratios for quasi-circles. Any random fraction that the quotient starts with 3.14 represents a quasi-circle ratio.

Example of quasi-circle ratio:
3.1428571 in the fraction above

(Please note that capital C, for the purpose of this theory, indicates automatically that the circumference of a quasi-circle is an approximation of the circumference of a perfect circle.)

The following will help the reader to identify quasi-circle fractions in accordance to the two criteria provided previously. The fraction 22/7 represents a quasi-circle ratio because it meets both required criteria that define approximate circles: it is a rational number expressed as a fraction made of positive integers, and its ratio starts with 3.14. But $10/3 = 3.333 \ldots$ is not a quasi-circle ratio because it only meets one of the two required criteria. It is a rational number that meets the first requirement, but it does not start with 3.14.

Pi, the approximate ratio of the perfect circle to its diameter, does not fall into the category of quasi-circles ratios even though it starts with $3.14 \ldots$ It fails to meet the first requirement because it is an irrational number and cannot be expressed as a fraction. *Pi is only used as a guide to calculate quasi-circle fractions.* The perfect circle follows the same mathematical formulae as quasi-circles, except that in the case of the perfect circle, the rational number that represents c/d or $\frac{1}{2}C/R$ is unknown.

Quasi-circle ratios (QCR) existed since the time of the Egyptians except they were simply not known in ancient times by such a name. Fractions expressing the ratio of a circumference to its diameter have been recorded since the time of the Egyptians and may go as far back as the Sumerians, but they were never expressed in terms of quasi-circle fractions. Old fractions such as 22/7 offered by Archimedes in 225 BC or 355/113 found in the works of the Chinese Zu Chongzi ca. 263 meet both criteria for quasi-circle ratios; their ratios start with 3.14, and they are both rational numbers made of positive integers. But the difference between a quasi-circle and a perfect circle was not established by the scholars as yet, neither mathematically or conceptually, and therefore could not have been explained by the ancient or modern mathematicians. This difference is being revealed for the first time in this work.

As this study will show, the world had not yet fully completed their study of fractions and ratios but moved on from fractions to decimals, leaving behind perhaps one of the most important concepts hidden in the midst of common fractions. The secret of the calculations of the inner and outer dimensions of the circle lies mostly in the quasi-circle fraction, not just in its equivalent ratio. These hidden concepts are encompassed in the *Fabius universal ratio formula* proposed by the author in the monad conjecture. They (the hidden concepts) could have been totally lost to mankind since decimal fractions had almost completely replaced common fractions two hundred years after Christoff Rudolff published his *Exempel-Büchlin* in 1530.

Once fractions were replaced by the decimal concept, it became even more difficult to find the true meaning of the universal ratio formula concept; and even if the true meanings were found, it does not mean that it would be understood how a ratio would relate to a geometric figure unless the researcher was doing some specific research in geometry. Modern researchers were certainly convinced beyond the shadow of a doubt that there was nothing left to know from the elementary knowledge of common fractions. Keep in mind that mathematicians were not fully persuaded of the incommensurability of pi until the German mathematician Carl Louis Ferdinand von Lindemann proved in 1882 that pi was transcendental and therefore, at the same time, proved the incommensurability of the dimensions of the perfect circle. Once pi was declared transcendental, it made the use of fractions equivalent to pi almost obsolete.

Objective F. Calculating quasi-circle fractions

Calculating quasi-circle fractions is as simple as converting decimals into fractions. Once the reader gets used to the idea, a quasi-circle perimeter and area can be calculated in a matter of minutes, if not seconds. Remember we can get a lot more information from the calculation of quasi-circles than from the calculation of the perfect circle. So far, I have delivered everything I promised, and it is only a matter of time before I surrender the remainder.

It is possible to calculate an infinite number of fractions representing the ratio of quasi-circles. As I stated before: *irrational numbers are just numbers waiting to be rationalized.* Various formulae can be used to convert decimals into fractions, and these can be found in many mathematics textbooks or on the Internet. Advanced calculators such as the TI-84 from Texas Instruments allow us to convert readily decimals into fractions.

Decimals can be converted into fractions according to their place value. One very simple method to convert decimals into fractions is to find the place value of the digit after the decimal point and use it as the denominator of the fraction. Fortunately, the place value can also be computed as the increasing values of 10^n. The first place value after the decimal point is equal to 10^1, second place value 10^2, third place value

10^3, etc . . . For example, if we want to convert pi into a fraction using one digit after the decimal point, we find the place value for the first digit after the decimal point and use it as a denominator:

Place value of first digit after the decimal point is:
10 x 1 = 10 or 10^1 and use 10 for the fraction's denominator for the first decimal digit

3.1 = 3 1/10 = 31/10

Place value of second digit is:
10 x 10 = 100 or 10^2 and use 100 as the fraction's denominator for the second decimal digit

3.14= 3 14/100 = 314/100 divided by 2 = 157/50

Place value of third digit is:
10 x 10 x 10 = 1000 or 10^3 and use 1,000 as the fraction's denominator for the third decimal digit

3.141= 3 141/1000 = 3141/1000

Place value of the fourth is: 10x10x10x10 = 10,000 or 10^4 and use 10,000 as the fraction's denominator

3.1415 = 3 1415/10000 divided by 5 = 6283/2000

Except 31/10, all the above fractions meet the criteria for quasi-circles. The fractions are reduced to their simplest expression, and fractions for pi can be calculated ad infinitum.

Hopefully now the reader gets the idea. Using this method, pi can be rationalized to any extent; and for each new digit added to pi, we can come up quickly with a new equivalent fraction. So there is a paradox in regards to irrational numbers—they can be rationalized as soon as they acquire existence. This makes all irrational numbers subject to infinite rationalization!

We can even use the formula $r \cdot \pi = \frac{1}{2} C$ and rounding the results to the nearest whole number to represent C. C symbolizes the approximate circumference of a circle. Since pi has already been calculated to over two trillion digits, pi can be used ad infinitum as a guide to compute fractions that represent approximate circles as long as their ratio meets the two criteria for approximate circles. This method, unlike the previous method, is empirical but can generate a whole new set of fractions for quasi-circles that are equally fun to work with.

For example:

r x π = $\frac{1}{2}$ C   **Rounded to nearest integer**

7 x 3.14=21.98   22/7

113 x 3.1415926 = 354.99996   355/113

As you can see, the ancient value of pi, 22/7, given by Archimedes and the Chinese value 355/113 given by Zu Chongzhi meet the two requirements that qualifies them as quasi-circle fractions.

Another method can also produce a new generation of geometric quasi-circle fractions by using the denominator in the form of 2^n and multiplying it by pi, and rounding it to the nearest whole number to obtain the numerator. Once the fraction meets the two quasi-circle ratio criteria, we have a fraction made of geometric dimensions because the key in computing quasi-circle ratios and fractions is based on the radius. The reader is invited to use his imagination to develop further this theory.

Using this method, I have computed quasi-circles fractions based on geometric dimensions. The following fractions represent quasi-circles with geometric dimensions because the denominator represents the geometric progression of 2^n:

3217/1024 = 3.1416015625
105414357/33554432 = 3.14159265160 . . .

The denominator of 1,024 is equal to 2^{10}, and the denominator of 33554432 is equal to *(2 exponent 25)*. These fractions are considered great measurable versions of pi (MVP).

Once we have found the fraction that represents the degree of accuracy that we need to achieve, we have all the elements necessary to calculate the dimensions of the quasi-circle with extreme precision.

This is the reason why I stated before that the function of pi has become much more extended than its previous role, and it has now become a guide in the calculation of quasi-circles. The author also reiterates what he stated before in the first chapter that pi needs the quasi-circle theory to achieve completeness, and the quasi-circle theory needs pi to reach fulfillment.

After calculating pi decimally, there is a need to revert back and find one equivalent fraction that represents pi to the extent that it has been decimally calculated. This fraction is fundamental to the quasi-circle theory and once more contains all the necessary elements to calculate the dimensions of quasi-circles.

Objective G and H. Calculating the Level of accuracy of quasi-circles and defining concentric and eccentric quasi-circles

The level of accuracy of a quasi-circle is equal to:
QCR - Pi

When we compare a quasi-circle ratio (QCR) to the approximate ratio of the perfect circle, the mathematical difference that we find is equal to the amount of imperfection of the quasi-circle relatively to the perfect circle, or simply the difference between the dimensions of a perfect circle and a quasi-circle. This degree of difference is called the level of accuracy of a quasi-circle. It is the equivalent of superposing an imperfect circle over a perfect circle to see the visual difference; but instead of looking at visual differences, we are looking at mathematical differences, which can be a lot more accurate than a visual difference. Imagine calculating a quasi-circle accurate to two trillion digits. This difference can only be perceptible through mathematics and cannot be detected by any known instruments. Of course, the expression of roundness for a circle must also be understood to fully understand these comparisons, and this is explained in another chapter.

Eccentric and concentric quasi-circles

Computing the level of accuracy of approximate circles led to two different types of quasi-circles: concentric and eccentric. If QCR-Pi is positive, we have an eccentric quasi-circle, meaning its ratio is slightly greater than pi; and if it is negative, we have a concentric quasi-circle, meaning its ratio is slightly smaller than pi.

The level of accuracy of a quasi-circle is determined by comparing quasi-circle ratios in Table 1 to pi, as illustrated in Table 2:

Table 1

C = Circumference
D = Diameter
QCR = Quasi-Circle Ratio

Once again, the new terminology being introduced is QCR, meaning quasi-circle ratio.

Examples of quasi-circle ratios:

½ C/R = QCR

$22/7$ = **3.1428571428** . . .
$3217/1024$ = **3.1416015625**
$355/113$ = **3.1415929203** . . .
$105414357/33554432$ = **3.1415926516** . . .

The true value of pi is used as a guide for comparison purposes: Pi = **3.1415926535** . . .

In the table above, each QCR digit written in bold letters matches a digit of pi and helps the reader to readily compare the value of pi against the QCR value. The reader can easily see that any desired degree of accuracy can be achieved to calculate the area or perimeter of a circle by using a quasi-circle ratio. In other words, we can always calculate a fraction representing a quasi-circle ratio beyond 2.7 trillion digits if necessary.

All the fractions above meet the quasi-circle definition because their ratio starts with 3.14 . . . and are made of rational numbers with positive integers. They contain all the necessary elements to calculate the dimensions of a quasi-circle.

Table 2

QCR–Pi = Level of Accuracy

3.1428571428 - 3.1415926535 = + 0.0012644893
3.1416015625 - 3.1415926535 = + 0.0000089090
3.1415929203 - 3.1415926535 = + 0.0000002668
3.1415926516 - 3.1415926535 = - 0.0000000019

Please note that the plus sign in the level of accuracy indicates an eccentric quasi-circle because it is greater than pi, and a minus sign indicates a concentric quasi-circle because it is smaller than pi. The type of quasi-circle, whether concentric or eccentric, does not affect its calculations.

Objective I. Old quasi-circle fractions have new meanings and function.

In the quasi-circle theory, a quasi-circle fraction reduced to its simplest expression is always equal to ½*C*/*R* and doubling the fraction would give the relationship of *C*/*D*. The fractions are different while the ratio stays the same.

Please note the capital *C* and the capital *R*. The reason for the big *C* and the big *R* is because they have a different meaning from the familiar *c*/*d* that expresses the relation between the circumference and the diameter of a circle. The capital *C* always signifies an approximate circumference relatively to the perfect circle's circumference. It also represents the number of rectified arc units that makes the perimeter of a quasi-circle. *R* represents the number of subunits in the radius or the divided base.

In the quasi-circle theory, we use a polygon concept to calculate the dimensions of the circle. For instance, if we consider the fraction 22/7 = 3.1428571; ½ C = 22 and R = 7, it means ½ C is equal to 22 *rectified arc units* for a circle of unit radius; but the divided base of the radius is parted into 7 *subunits*. Doubling the fraction gives the relation of C/D or 44/14 = 3.1428571 based on the number of arc units (44) that makes the circumference over the number of subunits (14) in the divided base.

When we use the denominator of a fraction that has been reduced to its simplest expression as ½C/R, when we double the fraction, the relationship of C/D *is always equal to an even number* and this will prevent the fractionalization of the monad when we try to compute the area of a circle. The quasi-circle theory will be integrated later in the monad conjecture, and we cannot afford to fractionalize the monad. Everything must fit together, and I have the purists on my side.

We were always using the ratio as an approximation of pi, and we had no use for the fraction itself. The ratio always worked, but we were never able to extract all the information that comes from the fraction itself. In the quasi-circle theory, we have as much use for the fraction as the ratio.

Objectives J to N.

Calculation of the area and perimeter of quasi-circles including their internal dimensions

When we use a quasi-circle fraction to calculate the area of a quasi-circle, we are using this fraction to modify the internal structure of a circle of unit radius in accordance to the concepts that was already established in the monad conjecture. The total area of the monads computed from the quasi-circle fraction is always equal to the ratio; thus, the ratio is equal to the area of a circle of unit radius.

The reader will remember that it was demonstrated in the monad conjecture that a square and a rectangle could be identified easily according to their ratio. It was not possible to show the reader how the circle could be added to this concept because he was not familiar with the quasi-circle theory as yet and could not technically understand how to identify a quasi-circle. The reader was simply not aware of the two criteria that had to be met in order to identify a quasi-circle according to its ratio.

A circle computed from a quasi-circle fraction represents a model circle of unit radius and reflects all the specifications that are particular to its ratio. Once we know the specifications of the model circle of unit radius, then circles of greater diameter can be computed by using almost the same (πr^2) formula that we used to calculate circles. The only difference is instead of using π, we will use the quasi-circle ratio $(r^2 \cdot QCR)$, the ratio computed from the quasi-circle fraction in order to calculate circles of greater diameter. Once the area and the perimeter have been computed for greater circles, we can refer back to the monad found in the model circle of unit radius and use the ultimate square or the ultimate arc unit to compute precise measurement for greater circles, if needed. This is a concept that we never grasped before.

After all, we never had any serious problems computing circles of unit diameter or greater. We simply never understood conceptually how to fragment the internal structure of a circle of unit radius in order to find the specific dimensions inherent to a circle of a particular ratio.

Theoretically, the area of quasi-circles is made of squares ranging from minute to subatomic that can be arranged in a circular manner to create the illusion of a perfect circle. These minute squares that make the area of a quasi-circle are the monads. They have a specific size and specific dimension, and they represent the ultimate element of physical reality for a particular ratio. As we have seen before, in the monad theory the monads are squares that are inherent to a ratio, and the particularities of the ratio were merged or associated with a geometric figure.

These squares do not seem to require any further division in order to represent the geometric figure equivalent in area to their ratio. It is uncertain whether the tiny squares can be actually arranged into shapes that would allow them to mimic the shape of a perfect circle. It remains speculative if the monads contained in a ratio can be attributed the shape of a circle, even though theoretically, if the squares are small enough, they should be able to represent a circle. The only thing that is absolutely sure, at this point, is that the area of an approximate circle is exactly equal to the sum of the tiny squares contained in their ratio; and it has been demonstrated in our study of the monad conjecture that these monads can be arranged at least in one instance to represent a model square or a model rectangle in accordance with their equivalent ratios. The author also believes that the monads could also be arranged, at least in one instance, to represent the shape of a quasi-circle.

For the purpose of teaching more effectively, I have duplicated the Fabius universal ratio formula along with propositions 1 to 6 established previously for quasi-circles in order to make this reference more accessible to the reader, and also to remind him one more time that it is the same universal ratio formula invented by the author that is being modified to calculate a quasi-circle, just like we did before in the case of the model square and the model rectangle:

Fabius universal ratio formula:

$$a/b = (1/b)^2 \, (a \cdot b)$$

At a glance, here is the formula to calculate the area of quasi-circles. In this formula, we have changed a/b into $\frac{1}{2}C/R$; a has been changed to $\frac{1}{2}C$ and b has been changed to R.

The *modified Fabius universal ratio formula* to calculate the dimensions of the circle is:

$$\tfrac{1}{2}\,C/R = (1/R)^2 \, (\tfrac{1}{2}C \cdot R)$$

Proposition 1. The base of a model quasi-circle is always equal to unity, and the radius represents the base of quasi-circles. The divided base of a model quasi-circle is always equal to its dividing factor, R, and R always represents the number of subunits pertaining to the base of a model quasi-circle. The side of a single monad in a quasi-circle is equal to 1/R and also represents the ultimate element of measurement for a particular model quasi-circle, and this measurement cannot be further divided by definition. The number of subunits (R) multiplied by the length of the side of a single monad 1/R is equal to the base of the quasi-circle that is equal to unity (R·1/R).

Proposition 2: The ratio of a quasi-circle is always equal to 3.14 . . . (see objective D, rationalization behind using any ratio starting with 3.14 . . . as the ratio of all circles).

Proposition 3: The ratio of a circle of unit radius is always equal to its area or ½ its perimeter.

Propositions 4 to 6 will be demonstrated as we pursue our study of the monad conjecture after the reader gets fully acquainted with the quasi-circle theory.

Proposition 4: Model rectangles and model quasi-circles that have in common the same divided base and the same fraction and ratio are also equal in area.

Proposition 5: If a model rectangle and a model quasi-circle have in common the same divided base and the same fraction and ratio, their area is equal and the perimeter of the rectangle is equal to the perimeter of the circle plus its diameter.

Proposition 6: If a model rectangle and a model quasi-circle have in common the same divided base and the same fraction and ratio, their area is equal, and the perimeter of the circle is equal to the perimeter of the rectangle minus both of its widths.

Proposition 7: If a model rectangle and a model quasi-circle have in common the same fraction and ratio, the perimeter of the circle is equal to both sides that make the length of the rectangle, and its diameter is equal to both widths of the rectangle.

Example 1:

Problem

Compute the area and the perimeter of a circle of unit radius by using Archimedes's ratio $22/7 = 3.1428571$ *(please note that we only use the ratio of the fraction)*

Let's compare the current method to the new method:

Current method to calculate a circle of unit radius

Area of the circle:

$\pi r^2 = 3.1428571 \cdot 1^2 = 3.1428571$

Circumference of the circle:

$\pi \cdot d = 3.1428571 \cdot 2 = 6.2857142$

Comparing the two methods, we must ask the following questions for the current method being used:

1. What is the side of the ultimate square that makes the area of this circle of unit radius?
    Unknown!
2. What is the area of the ultimate square that makes the area of this circle of unit radius?
    Unknown!
3. What is the number of squares that makes the area of this circle of unit radius?
    Unknown!
4. What is the specific length of each rectified arc unit needed to make the perimeter of this circle of unit radius?
    Unknown!
5. What is the total number of rectified arc unit needed to make the perimeter of this circle?
    Unknown!
6. What is the method to verify that the ratio equals ½ the circumference of a circle of unit radius?
    Unknown!
7. What is the method to verify that the area of a circle of unit radius is equal to its ratio?
    Unknown!
    What is the expression of roundness for this particular circle?
    Unknown

New method:

Compute the specifications for a model circle of unit radius using a quasi-circle fraction such as $22/7 = 3.1428571$.

My intent is to get the reader familiarized with the new terminologies as we go along.

½ C/R = QCR
$22/7 = 3.1428571$

(In this fraction, QCR means quasi-circle ratio; e.g. QCR = 3.1428571)

A quasi-circle fraction reduced to its simplest expression is always equal to ½C/R and represents the core values of a circle of unit radius.

$22/7 = ½C/R$;

R = divided base or number of subunits of the radius equal to unity (e.g. in fraction 22/7, R=7)

C = Total number of rectified arc units needed to make the circumference of a Quasi-circle (e.g. in fraction 22/7, ½C = 22)

Based on the above facts, let's compute the specifications for a quasi-circle of unit radius using the quasi-circle fraction $22/7 = 3.1428571$:

Explanation of the Modifed Fabius Formula to calculate the dimensions of the circle:

½ C/R = (1/R)² (1/2 C · R)

Objective J. Formula to determine the side of the monad is: **1/R**

1/R $= 1/7 = 0.1428571$ side of monad

Objective K. Formula to determine the area of the monad is: (1/R)²

(1/R)² $= (1/7)^2 = (0.1428571)^2 = 0.020408163265$

Objective L. Formula to determine the number of squares that makes the area of this circle of unit radius is: ½ C· R

½ *C· R* = 22·7 = 154

Objective M. Formula to determine the area of a quasi-circle of unit radius according to its ratio for fraction 22/7 is: $(1/R)^2$ (1/2 C · R)

Fraction 22/7= ½C/R

$(1/R)^2$ (1/2 C · R)

$(0.1428571)^2$ (154)

$0.020408163265 \cdot 154 = 3.1428571$

Note: the area of a model circle of unit radius is equal to its ratio.

The *modified Fabius universal ratio* formula to calculate the dimensions of the circle at a glance is:

½ C/R = $(1/R)^2$ (1/2 *C · R*)

(in the fraction 22/7; ½ C = 22 and R =7)

22/7 =$(1/R)^2$ (1/2 C · R)

$22/7 = (0.1428571)^2$ (22 · 7)

$22/7 = 0.020408163265 \cdot 154 = 3.1428571$

Objective N. Equation to validate the accuracy of the area of a model quasi-circle against its ratio is:

$(1/R)^2$ (1/2 C · R) = QCR (Quasi-circle ratio)

$(0.1428571\ )^2$ (154) = QCR

$0.020408163265 \cdot 154 = 3.1428571$

So the area of a quasi-circle of unit radius for a quasi-circle fraction 22/7 = 3.1428571 has been validated against its ratio, and the area is equal to the ratio. The formula to calculate the area and the ratio are the same.

Objectives O to R. Calculating the perimeter of a quasi-circle of unit radius for a quasi-circle fraction = 22/7 = 3.1428571:

Example 1:

Objective O. Formula to find the number of arc units that makes the circumference of a circle of unit radius is: 2(1/2C) = C. (Please realize that the number of rectified arc units is not equal to the perimeter of a model circle of unit radius).

2 (1/2C) =C
2 (22) = 44 (44 rectified arc units)

Objective P. The formula to calculate the length of each rectified individual arc unit is the same as calculating the side of a monad:

1/R = *side of the monad*

1/7 = 0.1428571 (The side of the monad is also the length of each rectified individual arc unit needed to make the circumference of the circle)

Objective Q. Formula to calculate the perimeter of a quasi-circle of unit radius is:

(1/R) C = 0.1428571 · 44 = 6.2857124

This formula verifies that the number of arc units multiplied by the length of the side of an individual monad is equal to the perimeter of a circle of unit radius for quasi-circle fraction 22/7.

Objective R. Equation to verify that the ratio is indeed equal to ½ the circumference for quasi-circle fraction 22/7=3.1428571:

½ C = 22 or number of rectified arc units needed to make ½ the circumference for this model circle of unit radius.

The formula to verify ½ the perimeter of a circle against its ratio is equal to the number of units that makes ½ the circumference of a Quasi-circle multiplied by the side of the monad:

½ C (1/R) = QCR (Quasi-circle ratio)

$22 \times 0.1428571428 = 3.1428571$

The ratio has been verified against ½ the perimeter as promised in the first chapter.

Example 2:

Calculating the internal and outer dimensions of a circle of unit radius by using the same formula for fraction 355/113= 3.1415929203:

R = divided base or number of subunits of the radius equal to unity

C = Total number of rectified arc units needed to make a circumference of unit radius

QCR = Quasi-circle ratio

A quasi-circle fraction reduced to its simplest expression is always equal to ½C/R)

½C/R = QCR
$355/113 = 3.1415929203$

Formula to find the side of a monad is:
$1/R = 1/113 = 0.008849557522123$

Formula to find the area of a monad is:

$(1/R)^2 = (1/113)^2 = 0.00007831466833737$

Formula to find the number of monads that form the area of this quasi-circle is:

$\frac{1}{2} C \cdot R = 355 \cdot 113 = 40115$

Formula to calculate the area of this model quasi-circle of unit radius is:

$(1/R)^2 (1/2C \cdot R)$

$(0.008849557522123)^2 (355 \cdot 113)$

$0.00007831466833737 \cdot 40115 = 3.14159203$

Verifying the area against the ratio

The *modified Fabius universal ratio formula* to verify the area against the ratio at a glance is:

$(1/R)^2 (1/2C \cdot R) = QCR$

$(1/113)^2 (355 \cdot 113) = QCR$ (Quasi-circle ratio)

$(0.008849557522123)^2 \cdot 40115 = QCR$

$0.00007831466833737 \cdot 40115 = 3.1415929 \ldots$

The area of a model circle of unit radius for quasi-circle fraction 355/113 is verified against its ratio. The formula to calculate the area is the same as the formula to calculate the ratio.

Calculating the perimeter:

Formula to find the number of arc units that makes the perimeter is:

$2 (\frac{1}{2} C) = C$
$2 \cdot 355 = 710$ rectified arc units

The formula to calculate the perimeter of a model quasi-circle of unit radius is:

(1/R) C

(1/113) 710
(0.008849557522123) 710 = 6.283185841

Verification of ½ the perimeter against the ratio for a fraction 355/113 = 3.1415929 . . .

½ C (1/R) = QCR

355· 0.008849557522123 = 3.1415929203

In each case, the area and the perimeter have been validated against the ratio for a circle of unit radius relatively to the value of their respective quasi-circle fraction.

Objective S. Expression of roundness of a quasi-circle based on the old polygon concepts of Archimedes

Since quasi-circles have now entered the picture in mathematics, then we must also consider the criteria that would make one circle rounder then another one.

First, we must consider that quasi-circles make the use of the old polygon concept used by Archimedes long ago, and there is no doubt that a quasi-circle is viewed as an *n* side polygon in the quasi-circle theory.

Second, we must consider the level of accuracy of the quasi-circle by comparing the ratio of the quasi-circle to pi to determine the difference, but the ratio alone is far from being sufficient. The more sides the polygon contains and the smaller the size of the sides, the smoother the edge of the circumference will be. Remember the picture of the moon I previously showed to the reader. This picture shows that the moon appears to have a smooth rounded edge, but a powerful telescope will indeed show that the moon is made of peaks and valleys just like Earth.

Third, another factor is the size of the length of the side of the monad that makes the circumference of a circle. The smaller the side of the monad, the more sides we will have that make the circumference of the circle and the smoother the edge of the circle will be.

Fourth, the length of the monad must be considered in relation to the length of the circle's radius since it is the ultimate unit of length for the radius and the circumference.

Considering those four factors, the author was able to develop a formula to express the degree of roundness for a quasi-circle.

First, we calculate the length of the side of the monad that represents the ultimate measurement for the perimeter of a quasi-circle of a given ratio.

Let's compare two circles to see how the comparison method works:

½ C/R = QCR

355/113 = 3.1415929

3217/1024 = 3.1416015625

Using these criteria and comparing the ratio of these Quasi-circles ratio to pi, which is equal to 3.1415926535, we realize that the fraction 355/113 is a lot closer to pi than 3217/1024 and therefore more accurate.

However, when we look at the number of sides of the polygons 3217 representing ½ the circumference for a quasi-circle of unit radius with a divided base of 1024, and 355 as the circumference for a quasi-circle of unit radius with a divided base of 113, we realize that the number of sides for the quasi-circle fraction 3217/1024 is approximately 9 times greater than those of the circle equal to 355/113 by dividing 3217/355. Therefore, the circle with 3217 sides could appear as having a smoother edge than the one with 355 sides. So to have a better comparison of roundness, we can multiply the numerator and the denominator of 355/113 by 9 to obtain a better comparison of roundness. The new fraction 3195/1017 = 3.14159292 has the same ratio than fraction 355/113, but now we have a lot more sides to compare with fraction 3217/1024. This difference will also alter the dimensions of the side of the monad that makes this circle.

Now let's compare the sides of the monad:

For fraction 355/113, the side of the monad is:
1/113 = 0.00884955752213

For fraction 3217/1024, the side of the monad is:
1/1024 = 0.0009765625

The monad that makes ½ the circumference of the circle having for ratio 3217/1024 is smaller, but we can multiply the denominator and the numerator of 355/113 by n, at will, to obtain a larger fraction to make the edge of that circumference smoother without changing the ratio.

Now when we multiply the numerator and the denominator of fraction 355/113 by 9 and obtain a new fraction 3195/1017, the polygon that forms this quasi-circle has a lot more sides while conserving the same ratio. This new fraction is closer to the fraction 3217/1024, while having the same ratio as 355/113. The side of the monad for fraction 3195/1017 is (1/1017 = 0.0009832841) and only slightly larger than the side of the monad for fraction 3217/1024 (1/1024 = 0.0009765625). By bringing closer, the two fractions, we have a better method of comparison for the circumference's smoothness, while keeping the same ratio. Remember, the smoothness of the circumference depends on how many times we can fit the side of a monad into a circumference.

The side of the monad for fraction 3195/1017 is:
1/1017= 0.0009832841

So in order to ensure that we are getting a circle with a more accurate ratio and a better expression of roundness than fraction 3217/1024, we would have to multiply 355/113 by 10 to get a smoother circumference edge and more accuracy or 3550/1330 = 3.1415929 in order to beat the fraction 3217/1024 = 3.1416015625 which has a less accurate ratio compared to pi, and carries a monad relatively larger and has now less sides.

So the level of accuracy of the circle comes down to the monad, and the monad can also be compared to the divided base of the radius. The monad is the ultimate unit that makes the dimensions of the circumference of a circle and also the ultimate unit that makes the radius of a circle. By comparing the side of the monad to the length of the divided base of the radius, we should obtain the ratio of the level of roundness for a quasi-circle.

The ratio of smoothness of a quasi-circle can be calculated, first by finding the side of the monad, and second by dividing the side of the monad to the divided base of the radius that is equal to unity.

For example, let's find the side of the monad for Quasi-circle fraction 3217/1024

$\frac{1}{2} C/ R = 3217/1024$

Side of the monad is:

$1/R = 0.0009765625$

Ratio of roundness for the circumference of this quasi-circle is:

R = divided base of circle of unit radius

$(1/R)/R$

(1/1024)/1024

0.0009765625/1024 = 0.00000095367431640625

Incidentally, this ratio is the same as $1/R^2$:

$1/R^2 = 1/1024^2 = 0.00000095367431640625$

Therefore, the ratio of the side of the monad compared to the divided base of the radius is the same as $1/R^2$

So the formula to express the roundness of a circle can be expressed as 1 divided by the square of the "divided base" of the radius or $1/R^2$ which is equal to the area of the square of a single monad particular to a specific ratio:

For fraction 355/113, the formula for the expression of roundness is: $1/R^2$

R always represent the divided base of a quasi-circle of unit radius in the expression ½C / R and is always represented by the denominator of the fraction

For fraction 355/113
$1/113^2 = 1/12769 = 0.0000783146683373$

For fraction 3217/1024:
$1/1024^2 = 1/1048576 = 0.00000095367431640625$

For fraction 3195/1017:
$1/1017^2 = 1/1034289 = 0.000000966847775$

For fraction 3550/1330:
$1/1330^2 = 1/1768900 = 0.0000005653230821$

So the formula to express roundness works, and it is the area of the monad that determines the level of roundness of the circle, and it makes sense. The larger the denominator of the fraction, the smaller the monad will be; and the smaller the monad, the smoother the edge of the circle will be. The dimensions of the side of the monad are relative to the dimensions of the parted radius and the number of arc units found in the circumference of a circle.

None of the previous methods offered by the scholars to compute the area of a circle come even close to the enormous quantity of information that can be obtained by using the quasi-circle theory, and even more can be obtained by using the monad conjecture.

The new methods of computation offered by the author are elementary notions of the circle that were unknown to us before the quasi-circle theory, and should we compare the results obtained from the two methods of calculations, we will soon realize that we are able to obtain a lot more information with our new methods of computation

Formula to calculate the area of all quasi-circles according to the quasi-circle theory:

$r^2 \cdot$ *QCR (quasi-circle ratio); same as* πr^2

We have learned to calculate the dimensions of a model quasi-circle of unit radius. Now we can calculate circles of greater radius, just by using the ratio, like we always did before. Instead of using pi, we use the QCR (quasi-circle ratio).

Let's calculate the dimensions of a quasi-circle of radius = 3:

First, we calculate the quasi-circle ratio:
For fraction 22/7;

$\frac{1}{2}C / R = QCR$ *(quasi-circle ratio)*
$22/7 = 3.1428571$
$QCR = 3.1428571$

Second, we calculate the area of the circle based on the formula $r^2 \cdot QCR$ which is the equivalent of the current formula πr^2:

$r^2 \cdot QCR$ = area of circle (QCR = quasi-cirle ratio)
$3^2 \cdot 3.142857142857$
$9 \cdot 3.142857142857 = 28.285714285714$

If we want to know the number of monads in this circle, we can just divide 28.28571 by the area of a monad from this particular fraction and ratio:

Area of monad is:

$(1/R)^2 =$
$(1/7)^2$
$(0.142857142857)^2 = 0.020408163265$
or $1/R^2 =$
$1/49 = 0.020408163265$

To find the number of monads in the circle of radius 3 above, simply divide the area of the circle by a single monad:

28.285714285714/0.020408163265 = 1386

It takes 1386 monads with an individual area 0.020408163265 to make the area of a circle of radius 3; or [r² (½ C • R)] (1/R)²

Formula to calculate the circumference of quasi-circles is:

d · QCR
QCR = 3.1428571 (has already been calculated for fraction 22/7)

Let's calculate the perimeter of a circle of radius 3:
6 · 3.14285714285714 = 18.857142857142

If we want to calculate the number of arc units that makes this circumference, we can just divide 18.857142857142 by the side of the monad:

Side of monad:

1/R
1/7= 0.142857142857

18.857142857142/0.1428571422857 = 132 arc units
or in the fraction:
½ C/R = QCR
22/7 =3.142857142857; if ½ C = 22 so C = 44

So the formula to find the number of arc units for a circle of radius 3 could also be expressed as:

C • r
44 · 3 = 132 arc units

So, it takes 132 arc units multiplied by the side of a single monad equal to 0.142857142857 to make the circumference of a circle of radius 3.

As the reader can see, it is easy to calculate the inner and outer dimensions of circles of radius >1 by just using the ratio and then refer back to the dimensions of the model circle.

Back to the monad conjecture

Objective 11. When we concluded the chapter in objective 7, I promised to the reader that we would return to pursue our study of the monad conjecture in objective 9 to show that we have calculated three basic geometric figures according to their ratio. A model square, a model rectangle, and a model quasi-circle have been now calculated according to their ratio.

Objective 12. Before continuing with the monad conjecture, let's summarize what we've seen in objective 7. To keep it simple, when we consider a fraction a/b=q, think of a ratio (q) as a square if its ratio is equal to 1, or a rectangle if its ratio is different from 1, and a rectangle or an approximate circle if the ratio is equal to 3.14 . . . .

The reader is now very well acquainted with the quasi-circle theory, and we can proceed with our studies of the monad conjecture and demonstrate that all quasi-circles are equal to a rectangle of the same ratio, and therefore, the demonstration of the following theorem for proposition 4,5,6,and 7 is now overdue:

Proposition 4: Model rectangles and model quasi-circles having in common the same fraction and the same ratio are equal in area, and the individual monads that make their area are also equal.

Proposition 5: If a model rectangle and a model quasi-circle have in common the same fraction and the same ratio, the perimeter of the circle is equal to the perimeter of the rectangle minus both of its widths, and the widths of the rectangle are equal to the diameter of the circle.

Proposition 6: If a model rectangle and a model quasi-circle have in common the same fraction and ratio, the perimeter of the rectangle is

*equal to the perimeter of the circle plus its diameter. Thus, the diameter
is equal to both widths of the rectangle.*

*Proposition 7: If a model rectangle and a model quasi-circle have in
common the same fraction and ratio, the perimeter of the circle is equal
to both sides that make the length of the rectangle, and its diameter is
equal to both widths of the rectangle.*

*All rectangles can be geometrically squared; therefore, all quasi-circles
can be squared. Methods to square a rectangle can be found in many
textbooks that treat the subject of geometry or on the web and will not
be discussed in this work. (Squaring the circle in this manner could
satisfy a number of mathematicians, but the author is dissatisfied with
this solution because the monads that are equal to the ratio cannot be
fragmented by definition)*

We would not have known this fact unless we studied the monad
conjecture. Now I will bring back some of the basics taught earlier in
the monad conjecture in objective 7 in order to prove this theorem.

It has already been established in the study of the monad conjecture that
the ratio of the sides of a rectangle is always different from 1, when we
divide its base by its height (h/b). In the study of the quasi-circle, we
have established that a quasi-circle ratio starts with 3.14 . . . if we divide
½ its circumference by its radius. The author believes there is at least
one instance where the ratio could represent both geometric figures at
the same time, and it is when the ratios of both geometric figures are
equal. This is a rare opportunity offered only in the monad conjecture
to compare the area of a rectangle to a quasi-circle, because the monad
conjecture allows us to compute geometric figures in accordance to their
ratio. So, all we have to do is find a fraction with a ratio that is common
to a rectangle and a quasi-circle at the same time, and all quasi-circle
fractions meet those requirements.

Let's calculate the dimensions of a model rectangle of unit base in
accordance to the Fabius universal ratio formula by using a fraction
equal to 3217/1024 = 3.1416015625: (Please note that the ratio starts
with 3.14 . . . and is also a rational number represented by a fraction,
and therefore meets the two requirements of a quasi-circle fraction. It is

also different from 1, which is the only criterion for a ratio to represent a rectangle.)

$a/b = (1/b)^2 (a \cdot b)$

Let's replace a by h (height) and b remains b (base) to represent the dimensions of a rectangle of unit base by modifying the Fabius universal ratio formula:

$h/b=(1/b)^2 (h \cdot b)$

The ratio of the sides of the rectangle is $3217/1024 = 3.1416015625$. This formula will convert this fraction into the dimensions of a model rectangle of unit base as discussed before in the monad conjecture:

Calculating the dimensions of the rectangle according to the modified Fabius formula above:

$h/b = 3217/1024$

h = the number of units that makes the height of the rectangle

b = the number of units that makes the base of the rectangle or divided base

$1/b = 1/1024 = 0.0009765625$ or side of monad

$(1/b)^2 = (1/1024)^2 = 0.\,00000095367431640625$ or area of each individual monad

$(h \cdot b) = (3217 \cdot 1024) = 3,294,208$ or number of monads in this rectangle

$(1/b)^2 (h \cdot b) = 0.00000095367431640625 \cdot 3,294,208 = 3.1416015625$ or area of rectangle

Formula for the perimeter of a rectangle of unit base is:

h = *number of units that makes the height of the rectangle*

b = *number of units that makes the base of the rectangle or divided base*

Sides of rectangle is equal to number of units multiplied by side of monad; h · 1/b and b ·1/b

[(2 *h*) (1/*b*)] +[(2*b*) (1/*b*)]

[(2 · 3217) (0.0009765625)]+ [(2· 1024) (0.0009765625)] =

[6434 · 0.0009765625] + [2048 · 0.0009765625] =

6.283203125 + 2 = 8.283203125 (total perimeter of the rectangle)

The side of the model rectangle of unit base is:

Height = 6.283203125/2 = 3.1416015625

The base of the model rectangle equal to unity = 2/2= 1

The height and the base of the rectangle can be replaced by length and width, so we have a rectangle of width 1 and length of 3.1416015625.

Now let's calculate the dimensions of a circle of unit radius using a quasi-circle fraction equal to the same ratio as the previous rectangle 3217/1024.

Let's replace the Fabius universal ratio formula by the formula to calculate the circle:

(C= number of arc units that makes the circumference of a circle)

½*C/R = (1/R)² (1/2C · R)*

$(1/R)$ = 1/1024 = 0.0009765625 or side of monad

$(1/R)^2$=$(1/1024)^2$=0. 00000095367431640625 or area of each individual monad (individual square unit)

$(1/2C·R)$ = (3217· 1024) = 3,294,208 or number of monads in the circle

$(1/R)^2(1/2\,C \cdot R) = 0.00000095367431640625 \cdot 3{,}294{,}208 = 3.1416015625$
or area of circle

Perimeter of the circle is:

The formula to find the number of rectified arc units needed to compute the perimeter of this circle of unit radius:

$2(\frac{1}{2}C) = C$; C = number of arc units

$2(3217) = 6434$ arc units

The number of rectified arc units that makes the perimeter is 6,434.

The formula to calculate the perimeter of this quasi-circle of unit radius is:

$C \cdot (1/R) = 6434 \cdot 0.0009765625 = 6.283203125$

The diameter is equal to twice the number of the divided base of the radius multiplied by the side of the monad:

$2R \cdot (1/R) = 2048 \cdot 0.0009765625 = 2$

And the radius is equal to 1

Or $R\ (1/R) = 1024 \cdot 0.0009765625 = 1$

Verification of propositions 4 to 7 for model rectangle and model quasi-circle having the same fraction and the same ratio 3217/1024:

The side of their individual monad is the same:

0.0009765625

The area of their individual monad is the same:

0.00000095367431640625

The number of monad that makes their area is the same:

3,294,208

Their area is the same and their area is also equal to their ratio:

3.1416015625

The perimeter of the model rectangle minus both its widths is equal to the perimeter of the quasi-circle:

8.283203125 - 2 = 6.283203125

The perimeter of the model quasi-circle plus its diameter is equal to the perimeter of the rectangle.

6.283203125 + 2 = 8.283203125

The diameter of the model circle is equal to both widths of the model rectangle:

Diameter = 2 and both widths of the rectangle = 2

Both lengths of the model rectangle are equal to the circumference of the model circle:

3.1416015625 · 2 = 6.283203125 = circumference of model quasi-circle 6.283203125

To conclude this chapter, the author will confess that when he started his research eons ago, his aim was to square the circle, a problem that was declared impossible if we consider the transcendence of pi. Sometimes we are astonished by the results of our own research. The idea of quasi-circles never crossed my mind when I first started my research, but I was led to this theory based on the result of my search. I have not squared the perfect circle, but I have demonstrated that every quasi-circle is equal to a rectangle, and this process can be repeated ad infinitum.

Rectangles can be geometrically squared, so quasi-circles can be squared. As I mentioned before, methods to square a rectangle can be found in many geometry textbooks or on the web and will not be discussed further in this work.

At the end of the day, I realize that solving the problem of the quadrature of the circle would not have meant as much to me as my understanding of the monad conjecture and the quasi-circle theory. This research has been a great journey for me, and I hope I was able to meet the reader's expectation.

A friend of mine, Jim Hood, who has now passed away, said to me as we were having a drink one day, "The true test of a man's intelligence and knowledge is not what he does when he knows what to do. It is what he does when he does not know what to do."

I consider myself blessed by the Almighty to have accomplished this study in my lifetime and for being able to communicate it to others. It's been so much fun!

Conclusion

Now we have reached the end of our mathematical excursion. Some may argue that the author only calculated a rectangle that is equal in area to a circle. My contention is the following: even if a rectangle equal in area to a circle was calculated, we are able to calculate this rectangle for every quasi-circle ratio. We've made more progress in the study of the circle in this work than we have made in the past 2,500 years since the dimensions of quasi-circles can now be calculated according to their respective ratios. But we've done better than that. Our study of the monad conjecture indicates that when a rectangle and a quasi-circle of the same ratio are compared, the perimeter of the quasi-circle is equal to the perimeter of the rectangle minus both its widths, and the widths of the rectangle are equal to the diameter of the circle, proof that the dimensions of a circle were calculated beyond the shadow of a doubt. Furthermore, the total number of monads (individual squares) that makes the area of such a rectangle is also equal in area to a quasi-circle of the same ratio.

The quasi-circle theory could have stood its own ground and could have been published without the help of the study of the monad conjecture; but the proof is in the pudding, and we would not have been able to prove that a quasi-circle is equal to a rectangle of the same ratio without the study of the monad conjecture.

I have already suggested that a new quasi-circle is created with each new decimal digit added to the remainder of pi. I have also suggested that with each new digit added to pi, the dimensions of a quasi-circle get closer and closer to those of the perfect circle as our calculations approach infinity. I have reiterated time after time in this work that if pi is the approximate ratio of the circumference of the circle to its diameter, then the perimeter of the circle is also approximate. My work indicates that the true meaning of pi as a transcendental number was not well understood by the scholars before this work, and pi is conducive to an infinite number of quasi-circles but not to the existence of a perfect circle. Let it be clear that my work does not disprove the existence of a perfect circle; it only represents a study of the circle seen from a numerical point of view; an approach that never occurred to my predecessors.

I have also demonstrated that monads (individual squares) are an inherent part of every ratio and their sum is equal to the ratio. I have showed how the monads found in a ratio could be considered as random squares that can assume various shapes while conserving the same area.

At last, we all have to decide whether geometry is more accurate than numbers or numbers more accurate than geometry. Some scholars may argue that even without a compass, we all can imagine a perfect circle that all points are equidistant to its center. My contention is that the same scholars provided an approximate number (pi) as the ratio of the circle to its diameter, and this number is a result of their work, not mine. I was merely interpreting the meaning of their own numbers because they failed to understand the implications of their own calculations. These implications are what lead us to the quasi-circle theory.

In this work, irrational and transcendental numbers were demystified. The author showed that these numbers were greatly misunderstood, and in reality, irrational numbers are just numbers waiting to be rationalized or converted into a fraction. In fact, they could be subject to infinite rationalization. Furthermore, in practice, irrational numbers taken to any decimal limits are rational and have exactly the same effect as fractions from a geometric point of view; and at the end of the day, the circle will always be viewed as an n side polygon.

The author does not view the transcendence of pi as representing an imperfection of the perfect circle but rather as an independent artificial and irrational number that will always lack completeness.

Moreover, we must also assume that as we get closer to the truth, our mathematical techniques will also improve. I have calculated with precision the number of squares that makes the area of a quasi-circle and the number of rectified arc units that makes its circumference. I have verified the ratio against the area of a circle which radius is equal to unity, and I have verified the same ratio against ½ the perimeter of the same circle. I have provided a method to calculate quasi-circle to infinity. These are things that mankind was unable to achieve since Egyptian and Babylonian times. I have also provided a method of comparison to express a quasi-circle's roundness, a thought that was not expressed before by any of our predecessors.

It is difficult to put a value on information, but this work will also expand the consciousness of the world in regards to circles' concepts.

At last, it was not a lack of mathematical skills that was preventing us from calculating the inner dimensions of the circle; it was rather our lack of discernment and understanding of basic mathematics.

Dimensions of a rectangle equal in area to a quasi circle calculated according to the **Fabius formulae**.

Model rectangles and model quasi-circles of the same fraction and ratio are equal in area and the perimeter of the quasi-circle is equal to the perimeter of the rectangle minus both widths, and the widths of the rectangle are equal to the diameter of the quasi-circle.

Ratio of *sides* of model Rectangle 3217/1024= 3.1416015625

Ratio of model Quasi-circle $\frac{1}{2}C / R$ = 3217/ 1024 = 3.1416015625

Rectangle's inner and outer dimensions (*geometric dimensions*):

Side of monad = 0.0009765625
Area of monad =0.00000095367431640625
Length of rectangle = 3217 x 0.0009765625 = 3.1416015625
Width of rectangle (base) = 1024 x 0.0009765625 = 1
Number of monads in this ratio: 3,294,208
Area of rectangle: 3,294,208 x 0.00000095367431640625= 3.1416015625
Perimeter of rectangle = 2 + 6.283203125 = 8.283203125

(base should be at bottom)

Side L = length = 3.1416015625

A = 3.1416015625

W = width = 1; base
P = 8.283203125

Circle's inner and outer dimensions (*geometric dimensions*):
Side of Monad = 0.0009765625
Square of monad = 0.00000095367431640625
Radius of circle = 1024 x 0.0009765625 = 1 (same as width of rectangle)
Number of monads in this ratio: 3,294,208
Area of circle: 3,294,208 x 00000095367431640625 = 3.1416015625
½ Perimeter of circle = 3217 rectified arc units x 0.0009765625 = 3.1416015625
Perimeter of circle = 6434 rectified arc units x 0.0009765625 = 6.283203125

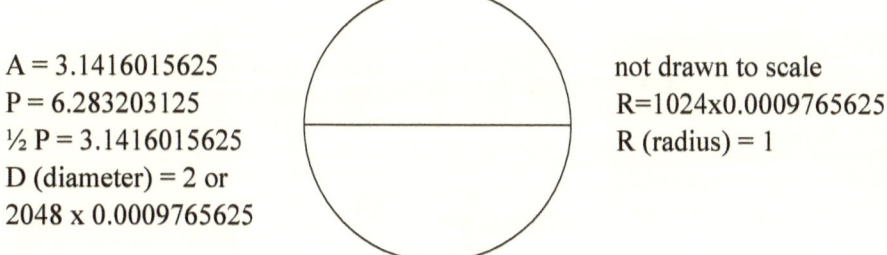

A = 3.1416015625
P = 6.283203125
½ P = 3.1416015625
D (diameter) = 2 or
2048 x 0.0009765625

not drawn to scale
R=1024x0.0009765625
R (radius) = 1

Perimeter of rectangle - (2 · Width) = perimeter of circle; both widths = diameter
8.283203125 - 2 = 6.283203125 or perimeter of circle; diameter = 2 x 1 = 2

Glossary

Approximate circles: Another name for Quasi-circles

Divided base: The number of units in the parted base of a model square, model rectangle, or a model quasi-circle, whose base is equal to unity. For a square and a rectangle, b is the divided base in the fraction h/b (height/base). For a quasi-circle, the radius is the base equal to 1 which is parted in a certain number of subunits equal to R, in the fraction $1/2C/R$. (Source is quasi-circle theory)

Monad: A square unit whose side is found by dividing the number 1 by the denominator of a fraction. Monads are indivisible by definition and represent the ultimate element that mimics physical reality for a specific ratio. For fraction a/b: $1/b$ represents the side of a monad, and $(1/b)^2$ is the area of the monad; b is always > 1.

Placed in a geometric context, a monad is the ultimate square unit that makes the area of a magnitude based on its ratio.

MVP: Measurable version of pi

Quasi-circle:

1. A circle in which the circumference is slightly disproportionate from its diameter.
2. A circle made of minute squares called monads that can be arranged in a circular shape theoretically to create the illusion of a perfect circle.
3. A circle with a smooth edge, based on the old polygon concept, that creates the illusion of a perfect circle.

Quasi-circle ratio: A ratio deriving from pi and originating from a fraction equal to $\frac{1}{2}C/R$ where the first three numbers of the quotient starts with 3.14 . . . and where R is > 1, and C and R are positive integers.

Pi taken to any extent beyond 3.14 . . . and represented by a fraction equal to $\frac{1}{2}C/R$.

Examples of quasi-circle ratios: 22/7 = 3.1428571 and 355/113 =3.14159

Quasi-circle theory:

1. The circle viewed from a numeric standpoint.
2. The quasi-circle theory is the study of approximate circles.

Ratio:

1. The relation of one proportion to another expressed as a fraction
2. The quotient of a fraction
3. A group of squares inherent to a ratio or quotient that can assume various shapes while conserving the same area; the ratio is always represented by a fraction a/b, and the denominator b of the fraction is always greater than 1.
4. The sum of the area of all the monads contained in a quotient or ratio

Reverse multiplication:

A term proposed by the author to indicate that the product resulting from a multiplication operation is smaller than the multiplicand, as in the case of proper fractions (e.g., $\frac{1}{4} \times \frac{1}{4} = 1/16$). As a reminder, a proper fraction is a fraction that the numerator is smaller than the denominator.

Technically for centuries, the scholars have applied the terminology "multiplication" erroneously to define the *reverse multiplication* that takes place in the case of proper fractions. According to the rules of multiplication, if we multiply a number such as $2 \times 3 = 6$, the multiplicand (2) can be added three times by the multiplier (3) to find the product (6), that is—2 + 2 +2 added three times is equal to 6.

But in the multiplication of proper fractions such as ¼ x ¼ = 1/16, the product of the numbers being multiplied move toward the infinitely small, as opposed to a multiplication of whole numbers where the product move toward the infinitely large; and the multiplicand cannot be added a certain number of times to find the product since the product is smaller than the multiplicand, proof that a multiplication was not performed in the case of proper fractions. It is obvious that the modus operandi for reverse multiplication does not follow the modus operandi of multiplication for whole numbers, and new criteria must be established in the case of reverse multiplication.

Therefore, in all humility, not only a new terminology is necessary in the case of multiplication of proper fractions; but also a new sign other than "x" or "·" may be needed to indicate a new operation. There is a great difference between operations resulting in a product that moves toward the infinitely large and one that moves toward the infinitely small. It is obvious that proper fractions do not follow the same rules of multiplication as whole numbers. We should provide and suggest new criteria for the reverse multiplication that takes place in the calculation of proper fractions.

Index

www.ingramcontent.com/pod-product-compliance
Lightning Source LLC
Chambersburg PA
CBHW030007190526
45157CB00014B/1005